<u>Disclaimer</u>

The publisher of this book is by no way associated with the National Institute of Standards and Technology (NIST). The NIST did not publish this book. It was published by 50 page publications under the public domain license.

50 Page Publications.

Book Title: Experimental Study of the Three Dimensional Internal Structure of Underventilated Compartment Fires in an ISO 9705 Room.

Book Author: Kelly M. Opert; Andrew J. Lock; Matthew F. Bundy; Erik L. Johnsson; Cheolhong Hwang; Anthony P. Hamins; Stephen P. Fuss; Ki-Yong Lee;

Book Abstract: This report documents a set of 9 full scale ISO 9705 room under-ventilated compartment fire experiments for the purpose of guiding the development of the National Institute of Standards and Technology (NIST) computer fire model - Fire Dynamics Simulator (FDS). The gas species composition and temperature throughout the interior of the compartment was mapped during quasi-steady burning conditions using movable measurement probes. In conjunction with the gas species and temperature measurements, global heat release rate, global burning mass rate, and local heat flux measurements were taken. The tests yielded detailed maps. From the data collected, the mixture fraction (with and without soot included in the calculations), local equivalence ratio, carbon monoxide and soot yields, fractional carbon monoxide and soot ratios, and combustion efficiency for each test were determined. Results from ethanol (a low sooting fuel) and heptane (a mildly sooting fuel) are presented. The results collected in this set of experiments were also compared and contrasted to the results of similar tests done in the previous report in this series of testing, NIST Technical Note 1603: Experimental Study of the Effects of Fuel Type, Fuel Distribution, and Vent Size on Full-Scale Underventilated Compartment Fires in an ISO 9705 Room.

Citation: NIST TN - 1736

Keywords: compartment fires; fire; room fires; heat release rate; soot; gas species; temperature; ISO 9705; heat flux; ethanol; heptane; carbon balance method; combustion efficiency; product yields; mixture fraction; local equivalence ratio; mass fraction; under-ventilated fires; ventilation-limited fires; liquid fuels; temperature; thermocouples

NIST Technical Note 1736

Experimental Study of the Three Dimensional Internal Structure of Underventilated Compartment Fires in an ISO 9705 Room

Kelly Opert
Andrew Lock
Matthew Bundy
Erik L. Johnsson
Cheolhong Hwang
Ki-Yong Lee
Anthony Hamins
Paul Fuss

National Institute of Standards and Technology
Technology Administration, U.S. Department of Commerce

NIST Technical Note 1736

Experimental Study of the Three Dimensional Internal Structure of Underventilated Compartment Fires in an ISO 9705 Room

Kelly Opert
Andrew Lock
Matthew Bundy
Erik L. Johnsson
Cheolhong Hwang
Ki-Yong Lee
Anthony Hamins
Paul Fuss

February 2012

U.S. Department of Commerce
John E. Bryson, Secretary

National Institute of Standards and Technology
Patrick D. Gallagher, Under Secretary of Commerce for Standards and Technology and Director

National Institute of Standards and Technology Technical Note 1736
Natl. Inst. Stand. Technol. Technical Note 1736, 92 pages (February 2012)
CODEN: NTNOEF

TABLE OF CONTENTS

LIST OF FIGURES

LIST OF TABLES

ABSTRACT

This report documents a set of 9 full scale ISO 9705 room under-ventilated compartment fire experiments. The gas species composition and temperature throughout the interior of the compartment were mapped during quasi-steady burning conditions. Particular focus is placed on minor carbonaceous gas species and soot. Fire protection engineers, fire researchers, regulatory authorities, fire service and law enforcement personnel use fire field models such as the National Institute of Standards and Technology (NIST) Fire Dynamics Simulator (FDS) for design and analysis of fire safety features in buildings and for post-fire reconstruction and forensic applications. These field models have historically showed limited ability to accurately and reliably predict the thermal conditions and chemical species in under-ventilated compartment fires. Among the various assumptions used in the development of previous versions of FDS, all chemical species were tied to the mixture fraction state relations. A single mixture fraction variable cannot be used for the prediction of carbon monoxide and soot, and the yield of these species was prescribed in FDS 4, rather than predicted. In fact, the yield of these species is usually not constant, but a complex function of their time-temperature history. While some previous studies have considered the mixture fraction to analyze experimental compartment fire data, few have considered minor hydrocarbon species and none have considered soot. Heptane and ethanol were burned in an ISO 9705 compartment with a 1/8 size door width (10 cm) in order to ensure under ventilated conditions. The fuels were sprayed into a 0.5 m^2 pan that was 0.1 m deep in order to maintain a steady heat release rate. This allowed for a long duration, quasi-steady state fire to be sustained. During this period, movable probes measuring temperature and gas species volume fractions were used to gather data at a multitude of locations. In conjunction with the gas species and temperature measurements, global heat release rate, global burning mass rate, and local heat flux measurements were taken. The tests yielded detailed maps. From the data collected, the mixture fraction (with and without soot included in the calculations), local equivalence ratio, carbon monoxide and soot yields, fractional carbon monoxide and soot ratios, and combustion efficiency for each test were determined. Results from ethanol (a low sooting fuel) and heptane (a mildly sooting fuel) are presented. The results collected in this set of experiments were also compared and contrasted to the results of similar tests done in the previous report in this series of testing, *NIST Technical Note 1603: Experimental Study of the Effects of Fuel Type, Fuel Distribution, and Vent Size on Full-Scale Underventilated Compartment Fires in an ISO 9705 Room.*

Keywords: Compartment fires; fire; room fires; heat release rate; soot; gas species; temperature; ISO 9705; heat flux; ethanol; heptane; toluene; carbon balance method; combustion efficiency; product yields; mixture fraction; local equivalence ratio; mass fraction; under-ventilated fires; ventilation-limited fires; liquid fuels; temperature; thermocouples

1 INTRODUCTION

This report describes new full-scale compartment fire experiments, which include local measurements of temperature, heat flux, species composition, and global measurements of heat release rate and mass burning rate. The measurements are unique to the compartment fire literature since they map the internal fire structure of the underventilated compartment. By design, the experiments provided a comprehensive and quantitative assessment of major and minor carbonaceous gaseous species at a number of locations within fires established in a full scale ISO 9705 room [1].

Fire protection engineers, fire researchers, regulatory authorities, fire service and law enforcement personnel use fire models such as the National Institute of Standards and Technology (NIST) Fire Dynamics Simulator (FDS) [2] for design and analysis of fire safety features in buildings and for post-fire reconstruction and forensic applications. Fire field models have historically showed limited ability to accurately and reliably predict the thermal conditions and chemical species in underventilated compartment fires. Formal validation efforts have shown that for well ventilated compartment fires, with the exception perhaps of soot, field models do quite well in predicting temperature and major species when experimental uncertainty is accounted for [2][3]. Inaccurate predictions of incomplete burning and soot levels impact calculations of radiative heat transfer, burning rates, and estimates of human tenability. High-quality (relatively low, quantified uncertainty) measurements of fire gas species, temperature, and soot from the interior of underventilated compartment fires are needed to guide the development and validation of improved fire field models.

The experimental results provided in this report are the continuation of a long-term NIST project to generate the data necessary to test our understanding of fire phenomena in enclosures and to guide the development and validation of field models by providing high quality experimental data. The experimental plan was designed in cooperation with developers of the NIST FDS model to ensure that the measurements would be of maximum value. Advanced development of FDS and other field models is extremely important, since it will lead to improved accuracy in the prediction of underventilated burning, typical of fire conditions that occur in structures. Improving models for under-ventilated burning will foster improved prediction of important life safety and fire dynamic phenomena, including fire spread, backdraft, flashover, and egress (involving the presence of toxic gas and smoke), which are critically important for application of fire models for fire safety.

1.1 Motivation and Objective

Field models, such as the FDS [2] are widely used by fire protection engineers to predict fire growth and smoke transport for practical engineering applications. Many field models numerically solve the conservation equations of mass, momentum and energy that govern low-speed, thermally-driven flows with an emphasis on smoke and heat transport from fires. Among the various assumptions used in the development of early versions of FDS, all chemical species were tied to a single mixture fraction variable by use of a set of mixture fraction state relations. A single mixture fraction variable cannot be used for the prediction of carbon monoxide and soot, and the yield of these species was prescribed in FDS 4, rather than predicted. In fact, the yield of these species is usually not constant, but a complex function of their time-temperature history. In practice, a knowledgeable user would attempt to pick yields that would reflect the anticipated ventilation condition of the simulation from literature values for well-ventilated burning, using data from a bench-scale apparatus, numerically predicted chemical equilibria [4], or from other sources such as the full scale experimental results presented here. Using this approach, the CO volume fraction for pool fire burning in an under-ventilated compartment can be underestimated by as much as a factor of ten.

FDS 5 [2] has included a simple predictive method for CO production. This revised method breaks the mixture fraction calculation into two parts resulting in a two-step chemistry model. This change in the chemistry of the model is an improvement over the prescriptive method used in FDS 4, however, it still over predicts CO substantially. A recent paper by the developers of FDS reported on the model validation of the reduced scale enclosure (RSE) experimental results [3]. They found that FDS 5 has improved its prediction of fires in this configuration. The worst agreement was observed with methanol, a very low sooting fuel. In general velocity and temperature data were well predicted from these experiments, with the exception of the largest fire sizes. The CO production model was improved substantially. However, there is still significant difference between the experiments and the model. As more soot was produced by the fuels and the fires became more underventilated, an under prediction of CO and an over prediction of CO_2 was observed. The authors attributed these effects to the specific assumptions made in the FDS CO prediction scheme.

In an effort to validate current fire models and to further the development of better predictive methods for fires, the current report presents new and unique data on the interior behavior of full-scale underventilated compartment fire experiments which builds on the previous data concerning RSE [5] and full-scale enclosure testing (FSE) focusing on the effects of fuel type, fuel distribution, and vent size [6]. The experiments are presented with analysis and experimental modeling results as a method of explaining the fire behavior and aiding in analysis.

1.2 Previous Work

Experimental research on enclosure fires has been on-going in fire research laboratories and academic institutions over the last 50 years. The motivation has varied from applied investigations studying particular fire scenarios to more fundamental work with the goal of understanding toxic species production behavior in fires. Some of the fundamental research that tried to ascertain ventilation and upper-layer effects on enclosure fire chemistry was conducted in well-controlled hoods. Sometimes, the main objectives of the research was to generally develop and validate fire models or particular structural fire simulations, while much of the research was conducted to acquire a better understanding of complex enclosure fire dynamics with a focus on chemical and thermal conditions. This section provides an overview of some of the recent research efforts in enclosure fires and highlights some of the more pertinent experimental work.

Research conducted at Harvard University and the California Institute of Technology in the 1980s explored fires burning under an exhaust hood to simulate the layer effect of an enclosure fire, e.g. [7-8]. The relative distance of the fire below the hood was adjusted to vary the entrainment of air into the plume before it entered the upper layer. These experiments focused on underventilated burning and the validity of the "global equivalence ratio" (GER) concept to correlate gas species in the upper layer. The GER is the fuel-to-air mass ratio normalized by the mass ratio required for stoichiometric burning. In a recent study, Brohez et al. explored the use

12

of a bench-scale calorimeter to measure fire properties of materials burning in underventilated conditions [9].

Research at NIST by Bryner et al. explored the global equivalence ratio concept and carbon monoxide production in a reduced (2/5) scale enclosure with natural gas as the principal fuel [10]. The results showed that the upper layer in enclosure fires is not homogeneous, and that CO can be produced in greater quantities than predicted by the GER concept, depending on temperatures and flow patterns developed within an enclosure. The subsequent effort [5] was meant to overlap some of the conditions explored by Bryner et al. and to repeat and fill gaps in the data. Pitts et al. expanded the work to full-scale and other fuels such as heptane and wood. It was established that wood pyrolysis in the upper layer of an enclosure fire can produce high concentrations of CO directly without further oxidation to CO_2 [11]. A subsequent study by Lattimer confirmed and expanded on this research [12].

Researchers at Virginia Tech investigated fires in a reduced-scale enclosure that directed the air inflow through slots in the floor connected to a duct where instrumentation was used to quantify air entrainment [13]. Several fuels were studied, and this configuration produced results consistent with GER predictions due to the more distinct, less dynamic nature of the gas layer structure. Later work used a more typical enclosure design and focused on transport of gas species outside the doorway and how it was affected by doorway geometry, soffit design, and hallway configuration [14]. More recently, Gann et al [15] conducted research on transport of toxic species in a full-scale enclosure with a corridor. These data were analyzed by Hirschler [16]. Researchers in Sweden conducted a study [17] of under-ventilated fires in an ISO 9705 room with a window vent of varying height. Several polymer fuel types were included in this study and measurements of local equivalence ratio and toxic gas species were performed.

Pitts [18] provided a comprehensive review of the application of the GER concept to predict CO concentration in building fires, using data from the Harvard and Cal Tech hood experiments [19][8], the Virginia Tech enclosure studies [13], and the NIST RSE experiments [10] . Several CO formation mechanisms were identified, which were substantiated by detailed chemical kinetic modeling. While the GER concept is of limited utility for predicting the local CO concentration, important aspects of enclosure fire dynamics and chemistry are highlighted in this paper.

Several recent experimental studies [20-22] have used very small scale enclosures (0.21 m^3, 0.06 m^3, and 0.05 m^3, respectively) while investigating under-ventilated burning of propane and heptane fires. These bench-scale studies described the structure and dynamics of under-ventilated burning including extinction, flame projection and flame stability. Another recent study [23,24] has used an intermediate-scale enclosure similar to that used for this paper, but a roof vent was added as well.

The first component of this research project focused on similar experimental measurements of a RSE [5]. The RSE was a 2/5 scale ISO 9705 room designed based on the previous studies of Bryner et al. [10]. Similar to Bryner et al.'s experiments, natural gas served as a fuel; the burning of heptane, toluene, methanol, ethanol, and polystyrene was also investigated. In most experiments, the fuel was controlled and metered by flow valves or pumped into a pool burner or spray nozzle. Experiments were run to near-steady conditions. Multiple fire sizes were run

13

consecutively to decrease the time required to approach steady-state. Ventilation was varied during some experiments by modifying the door opening. Two types of enclosure lining materials were investigated and compared.

In a later component of this project, natural gas, heptane, toluene, iso-propanol, polypropylene, nylon, and polystyrene were burned in a full-scale ISO 9705 room. The fuel was either allowed to burn freely in a pan or controlled and metered by flow valves or pumped into a pool burner or spray nozzle. Experiments were either run as free burns or at near-steady conditions. As in the first component of this test series, multiple fire sizes were run consecutively to decrease the time required to approach steady-state. The ventilation was varied during some experiments by modifying the size of the door opening. The data taken from these experiments were used to evaluate the effects of different fuel types, fuel distributions, and vent sizes. The results from these tests have been documented in NIST reports [5] [6].

Recently, NIST has conducted a number of high-profile case studies in which realistic-scale mock-ups of actual fire scenarios were recreated with the ultimate goal of improving building codes and standards. These studies included the World Trade Center disaster investigation [25] , the Rhode Island Station nightclub fire [26], and the Chicago Cook County Administration Building fire investigation [27]. The compartment fires in all of these studies burned real furnishings and became under-ventilated as the fire evolved. In addition, a series of large-scale compartment fire experiments were conducted to simulate an over-ventilated fire in a nuclear power plant cable room [28] to provide data for fire model validation.

1.3 Experimental Scope
While some previous studies have considered the mixture fraction to analyze experimental compartment fire data, few have considered minor hydrocarbon species and, with the exception of Ref. [5][6], none have considered soot. In tandem, accurate measurements of temperature at these same locations allowed analysis of thermal effects on species concentrations. Heptane and ethanol were used as fuels. The information gathered in the experiments presented here was used to map the gas species and temperatures of the interior of the underventilated compartment in three dimensional space.

The series of experiments reported here was conducted in a FSE. The enclosure defined in the international standard ISO 9705 "Full-scale room test for surface products" [1] is an important structure in which to conduct fire research since it is representative of a room in a residence and has been commonly used by other researchers. The experiments repeated and extended a part of the work of Bryner and coworkers [10] [18] as well as the authors previous work with a reduced scale enclosure [5] and full-scale enclosures [6]. The fuels were pumped at a steady rate and sprayed into a pan burner. The experiments were run at near-steady conditions. Multiple fires were run consecutively to keep the interior walls preheated, which in turn decreased the time required to approach steady-state. Ventilation remained constant for this set of experiments with a door width of 10 cm, 1/8 the size of the mandated ISO 9705 width. A 10 cm doorway was used for the experiments in order to force the room to reach under-ventilated conditions with a smaller fire size and therefore limiting the temperatures and thermal radiation within the room.

Heat feedback and natural ventilation strongly influence the structure and dynamics of the fire, such as the temperature field and the spatial distributions of combustion products. This study deliberately set out to investigate representative fire conditions at as many internal points as were practically possible. Two liquid fuels were tested, namely, heptane and ethanol. To allow for comparison, the ideal heat release rates of all of the fires were set to approximately 1000 kW. Combined gas species and temperature measurement probes were situated at different locations in the room and then moved throughout the room. Measurements were taken along the centerline and near one side wall at locations from near the floor to near the ceiling. The moving probes allowed for sampling in the upper layer, lower layer, and within the transition between the layers as well as near the ventilation opening, the burner, and the rear wall. Other measurements taken at stationary locations included oxygen, carbon dioxide, carbon monoxide, total hydrocarbons, and soot mass fractions, temperatures from two thermocouple arrays, and heat fluxes. Oxygen was measured with a paramagnetic analyzer and carbon dioxide and carbon monoxide were measured with non-dispersive infrared detectors. Hydrocarbons were measured with flame ionization detector (FID) analyzers. The quantification of hydrocarbon species was needed to describe the chemical structure of under-ventilated fires. Soot samples were extracted from within the enclosure and measured gravimetrically.

2 EXPERIMENTAL DESIGN

2.1 Design of the room

The ISO 9705 room [1] was used in this experiment due to its wide utilization in other works and to build upon the previous experiments with the RSE [5] and the FSE [6]. The RSE investigation looked at a variety of room construction materials and helped to guide the development of the final design of the ISO 9705 room which was used in the previous full-scale tests and for this set of experiments.

2.1.1 Dimensions

The FSE dimensions are illustrated in Figure 2.1. The design internal dimensions of the room were set to the ISO 9705 standard of 240 cm × 240 cm × 360 cm with a modified doorway of 10 cm × 200 cm. The floor of the enclosure was raised 35 cm above the ground. The height of the door was not varied. Due to the nature of the lining material and the fasteners used to hold it in place, there was some variability in the actual dimensions of the room. However, the as-built dimensions were measured extensively and uncertainty was found to be within ± 2 cm, well within the tolerance of the ISO 9705 standard of ± 5 cm. Additional measurements were taken periodically within the room between the experimental tests, and they never exceeded a uncertainty of ± 2 cm. A picture of the actual structure can be seen in Figure 2.2.

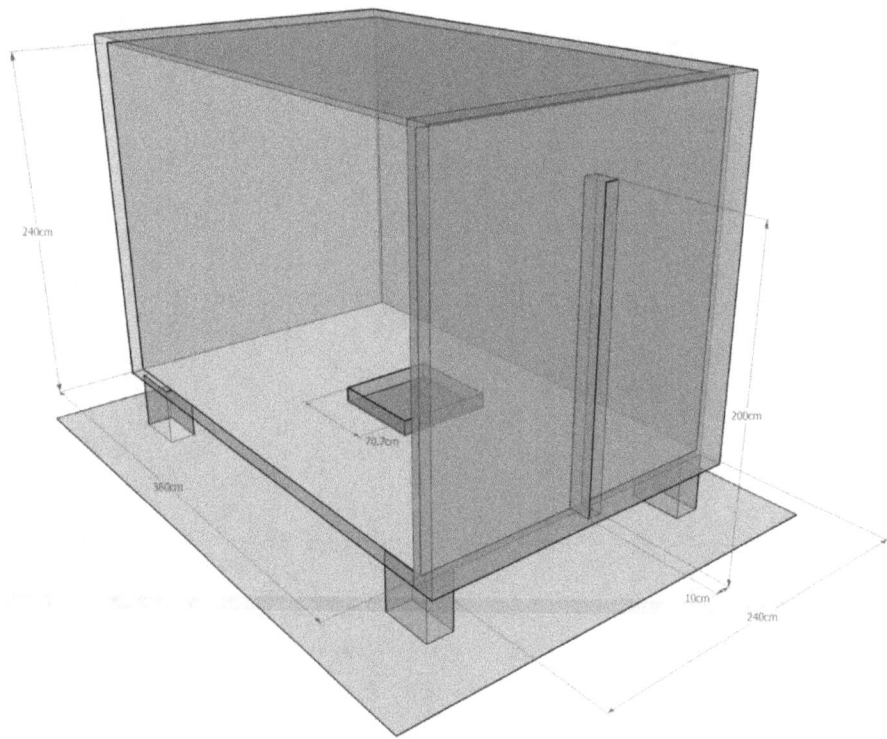

Figure 2.1 Internal dimensions of ISO 9705 enclosure used in these experiments with the altered 10 cm door width. All dimensions have an uncertainty of ± 2 cm.

Figure 2.2 Photograph of the actual ISO 9705 room used for experiments. The door inserts used to reduce the door width to 10 cm are displayed in this photograph.

2.1.2 Materials

The support structure of the room was built using 20 gauge (0.89 mm) steel structural studding and 20 gauge (0.89 mm thick) sheet steel. The floor of the structure was constructed of 0.48 cm thick steel sheet metal. The actual room used in the experiments can be seen in Figure 2.2. The studs and sheet metal were built such that their internal dimensions were 10 cm greater than the ISO 9705 standard. On top of the sheet metal (on the interior surfaces) were installed two layers of 2.5 cm thick, 128 kg/m^3 density, ceramic fiber blanket, K-lite^{tm} HTZ. The manufacturer's reported blanket composition was 30 % AL_2O_3, 54 % SiO_2, 16 % ZrO_2, and trace amounts of other components. The uncertainty in the composition of the primary four components is ± 1 %. The ceramic fiber blanket was held in place by alumina ceramic (99 % Al_2O_3) insulation retainers (Refractory Anchors Inc. model RA38) with a depth of 5 cm. These anchors are shown in Figure 2.3. The ceramic anchors were secured to the sheet steel wall with self-tapping sheet metal screws and washers. Insulation retainers were installed in the ceiling studs, spaced 40.5 cm, at 30.5 cm intervals along each stud. On the walls the insulation retainers were also installed with an arrangement of 40.5 cm by 30.5 cm near the top of the wall with the spacing increasing to 40.5 cm by 70 cm as the retainer placement approached the floor. Extra retainers were placed as necessary to hold edges and corners securely in place.

This structure design proved to be quite robust. Through a series of tests, only minor repairs to the blanket and ceramic retainers were necessary. The steel skin and steel studs held up well with the exception of the portions of the structure framing the doorway. In the vicinity of doorway the ceramic fiber insulation was wrapped around the doorway to protect it from the heat, and radiation from the room. This additional insulation was not sufficient to protect the studs from

17

excessive heat, causing them to soften and deform over time. This situation was further exacerbated by the convective heat transfer from the hot, fast moving gases leaving the enclosure.

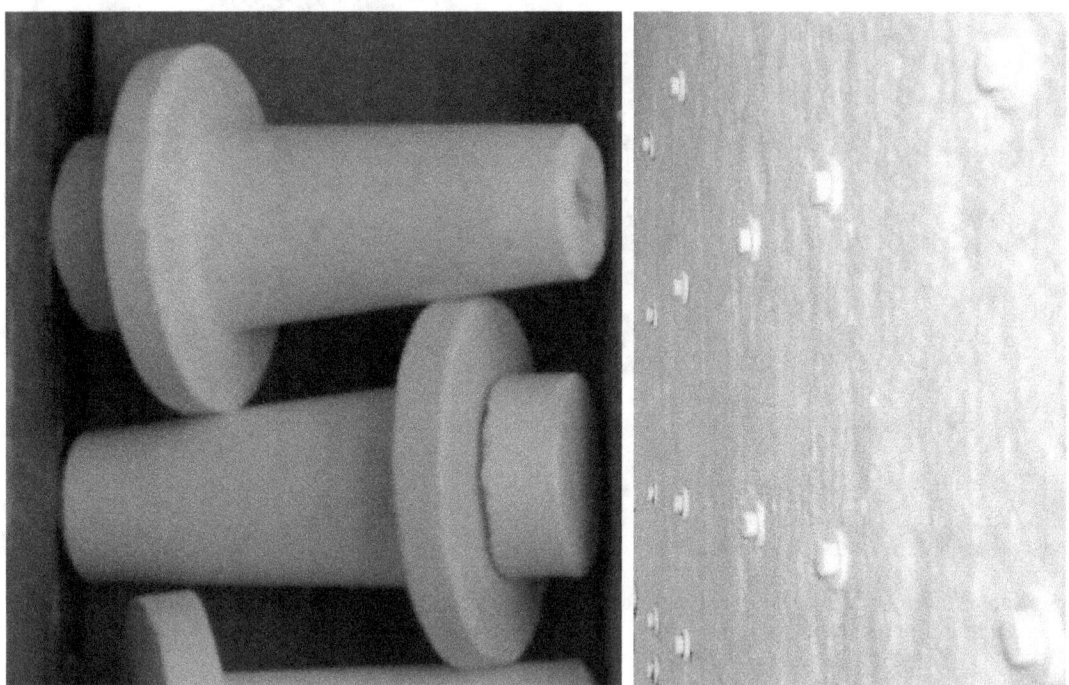

Figure 2.3 Ceramic insulation retainers used to secure the ceramic fiber blanket to the sheet steel walls. The actual retainer is shown (left) as well as its installed configuration (right).

2.1.3 Doorway Dimensions

A 10 cm wide doorway was used for the experiments. The height of each doorway was held constant at 200 cm. Inserts were constructed from steel studding and ceramic fiber blanket and installed to change the width of the door from 80 cm, the ISO 9705 standard, to 10 cm. The doorway inserts allowed doorway accessibility for work inside the room. Unfortunately, repeated heating and cooling of the doorway inserts resulted in deformations. Every attempt was made to ensure that the proper doorway sizing was maintained. The uncertainty was ± 10 % of the doorway width.

2.1.4 The Burner

One 70.7 cm × 70.7 cm × 10 cm pan, constructed of steel, 0.6 cm thick, placed in the geometric center of the room (Figure 2.1) was used with the spray burner configuration as seen in Figure 2.4. Different spray nozzles were utilized depending on the desired fuel flow rate. All spray nozzles were BETE Low Flow/Full Cone Whirl nozzles. BETE model number WL 1-1/2 was utilized in most experiments. The nozzle was constructed from stainless steel and featured a 90° cone angle. The pump flow rate was varied in order to provide different flow rates at the nozzle to produce different fire sizes. The pan was elevated 5 mm from the floor of the compartment and was positioned on top of a load cell. The load cell was utilized to measure the spray burner pan mass in addition to monitoring the pump flow-rate to determine if any fuel collected in the

18

pan. In this way, all of the fuel from the spray burner could be accounted for and used to calculate the overall combustion efficiency. Generally, the fuel did not accumulate.

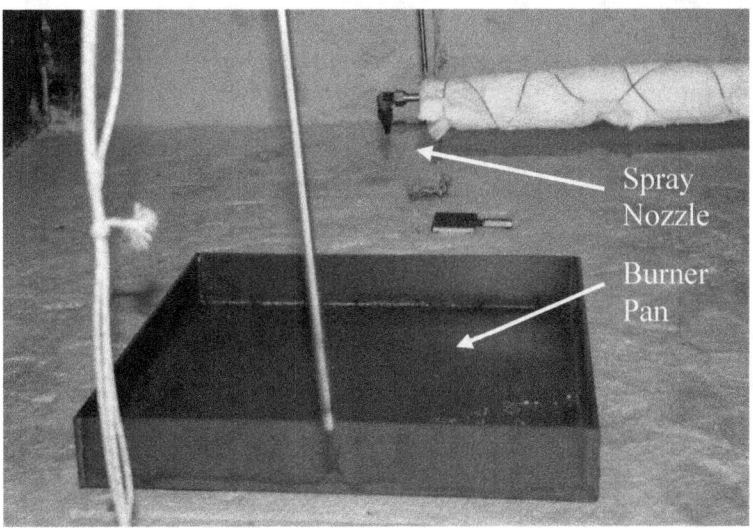

Figure 2.4: Picture of the spray nozzle and burner pan.

2.2 Overview of equipment

2.2.1 Calorimeter

Heat release rate (HRR) measurements were conducted using the 6 m × 6 m calorimeter at the NIST Large Fire Laboratory (LFL). The HRR measurement was based on the oxygen consumption calorimetry principle first proposed by Huggett [29]. This method assumes that a known amount of heat is released for each gram of oxygen consumed by a fire. The measurement of exhaust flow velocity and gas volume fractions (O_2, CO_2 and CO) were used to determine the HRR based on the formulation derived by Parker [30]. A detailed description of the methodology used for this measurement can be found in a previous report [31]. In 2001, the 6 m × 6 m square hood was installed in the LFL. A schematic drawing of the 6 m square hood is shown in Figure 2.5. The exhaust flow rate and extractive gas measurements were performed in a horizontal straight section of the 152 cm diameter duct on the roof of the LFL. The flow coefficient was determined using a natural gas burner to conduct a five point calibration before and after the test series. The flow calibration coefficients and uncertainties (± 2σ) for these tests ranged from 0.91 ± 0.04 to 0.93 ± 0.05. The calibrations were performed over a range of fire sizes from 500 kW to 3000 kW. The exhaust mass flow rate for the experiments described here varied from 12 kg/s to 17 kg/s.

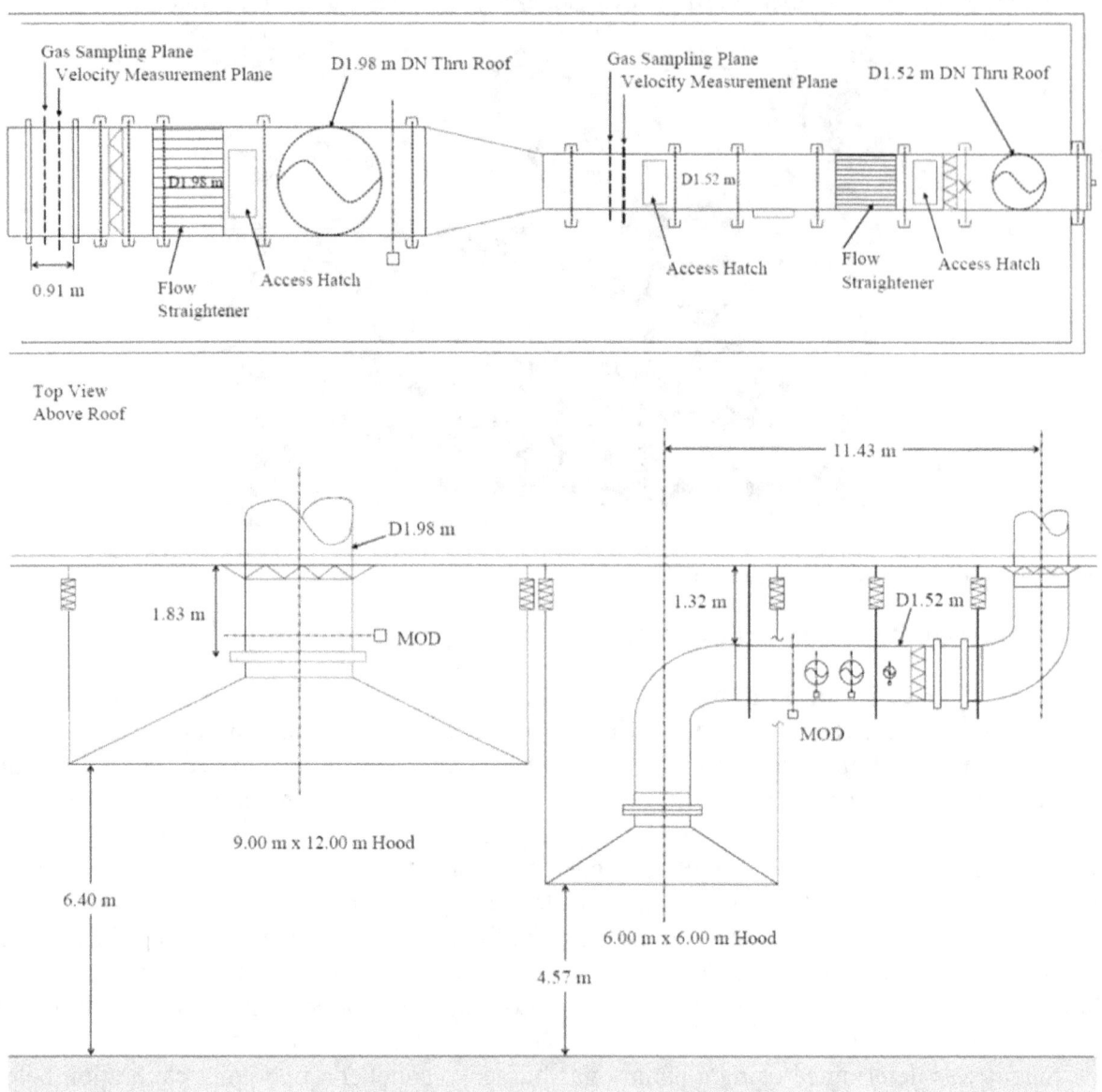

Figure 2.5: Schematic drawing of 6 m square hood and exhaust stack instrumented for calorimetry measurements. Taken from Ref. [31]

2.2.2 Gas Analyzers

Gas species and temperatures were measured throughout the FSE at multiple points during each experiment. Oxygen was measured using paramagnetic analyzers. The 10 % to 90 % response times (t_{10-90}) of the oxygen analyzers were less than 12 s. Carbon monoxide and carbon dioxide were measured using non-dispersive infrared (NDIR) analyzers. The t_{10-90} response times for the CO_2/CO analyzers were less than 5 s. Total hydrocarbons were measured using FIDs having a t_{10-90} response times of less than 1 s. The dried sample gas dew point temperature was measured using a thin polymer sensor which had a response time on the order of a minute. The total delay times for each of the analyzers were measured by initiating a small flame at the gas sample probe inlet for approximately 10 seconds and timing how long until a response was recorded by the gas analyzers. The delay times of each instrument did not significantly contribute to the uncertainty of the HRR measurement because multiple samples were taken throughout a pseudo-steady-state fire and combined. The response times were, however, used to correct the delay in the data.

The NDIR analyzers were spanned with a 3 % CO and 8 % CO_2 span gas. The three FID analyzers used in these experiments were designed to measure high volume fractions of hydrocarbons. The analyzers were factory calibrated for up to 50 % volume fraction of hydrocarbons as methane equivalent and were capable of measuring even higher concentrations. The primary span gas used for these meters was a 20 % volume fraction of methane with a balance of nitrogen. A span gas of 1 % methane in nitrogen was also used to periodically check the linearity of the detector. The FID burner fuel was 40 % hydrogen and 60 % nitrogen on a volumetric basis. The expanded (k = 2) relative uncertainty of each of the span gas volume fractions, including CH_4 (20 % CH_4, balance N_2), CO, CO_2 (3 % CO, 8 % CO_2, balance N_2), and O_2 (21 % O_2, balance N_2) was ± 1 %.

Each hydrocarbon analyzer had an internal filter to prevent soot from accumulating in the plumbing and internal sample pump which could lead to less sensitivity due to hydrocarbon contamination and also deterioration of some components of the instrument. It was later determined that additional external filtration of soot was necessary to protect the analyzers and enable a sufficient time period for sampling soot-laden flows. The external filter could be replaced much more frequently and easily than the internal filter.

Three sample probes were used to sample gas inside the enclosure at three locations simultaneously, and were moved vertically. After tests at one set of three points, the probes were moved to another lateral location, as discussed in section 2.3. The 2 m long probes were constructed of 0.95 cm (3/8 inch) nickel alloy tubing with an inner diameter of 0.78 cm (0.31 in.). An early experiment was executed using both cooled stainless steel tubing and uncooled nickel tubing. There was no measurable difference between the two types of tubing. Ultimately nickel alloy tubing was chosen for its higher temperature tolerance and to reduce the number of lines going to the movable probes.

A schematic of the gas sampling system can be seen in Figure 2.6. The total hydrocarbon analyzers were placed in the gas racks with the other analyzers. The gas sample stream water was removed with preferential diffusion membrane dryers consisting of a bundle of Nafion tubes purged with dry air to selectively remove moisture from the sample stream. The Nafion conditioner had no effect on most of the gas species of interest. However, polar organic compounds (i.e. ketones and alcohols) are trapped by the dryer. A large area filter was added between the heated line and gas dryer to collect soot. Because the external filters and transfer lines after the gas dryer were not heated, there was a potential loss of high molecular weight hydrocarbons due to condensation. Due to limitations in the flow capacity of the dryer, the gas analyzers were connected in series. A mass flow controller set to 1 L/min was used to control the flow through the $O_2/CO_2/CO$ analyzers. The flow to the hydrocarbon analyzer was split prior to the mass flow controller. A 5 way ball valve was connected to each analyzer to switch between the gas sample, zero calibration gas and span calibration gas. A dew point transducer was connected to the sample gas line prior to the oxygen analyzer. The oxygen analyzer had separate inlet ports for zero and span gases. A needle valve was used to set the total flow to 3 L/min (only a small fraction of this passed through the FID).

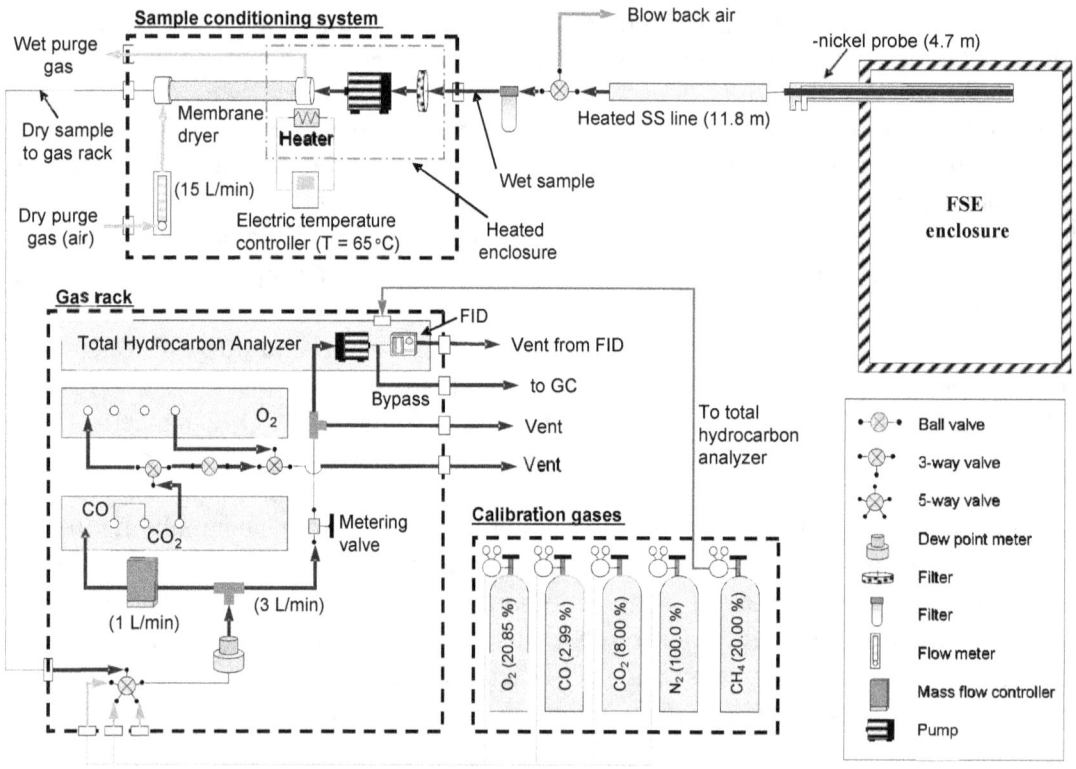

Figure 2.6: Schematic drawing of gas sampling system

22

2.2.3   Gravimetric Soot Sampling System

A gravimetric soot sampling system (shown in Figure 2.7) was used to measure soot mass fractions at two sample locations within the enclosure. The sample locations were chosen to build upon the previous RSE and FSE work. The design of the soot probe was similar to the gas sampling probes except that the inner diameter of the sample tube was 6.4 mm. The soot sampling probes were conditioned with 65 °C water flowing at 1.0 L/min. A three way solenoid valve was used to rapidly switch from a bypass to sample flow. A sample gas mass flow rate of 2.75 standard L/min (N_2 @ 0 °C, 101.3 kPa) was drawn through the collection filter for a period of 60 s to 300 s. The collection filter was a 47 mm round Zeflour membrane filter with an aerosol retention efficiency of 99.99 % for 2 μm sized particles. A gas correction factor was applied to the mass flow rate measurement to account for the gas composition in the enclosure. The amount of time for sampling was determined by monitoring the pressure drop across the filter to ensure an optimal amount of filter loading.

The collection filters (shown at the base of the probes in Figure 2.7 below) and probe cleaning pads were conditioned in a desiccant drier before and after the tests. After each soot sampling period, the probe was cleaned twice with gun cleaning pads. The total soot mass collected on both the filter and 2 cleaning pads was used in determining the soot mass fraction. Both the soot mass and sample mass flow rates were measured on a dry basis. For most of the tests conducted in this series between 10 mg and 200 mg of soot were collected during the 1 min to 5 min sample time. The extracted gas volume was corrected for the water removed. The conditioned filters were weighed using an analytic mass balance with an expanded uncertainty of 0.12 mg. The combined expanded relative uncertainty of the soot mass fraction measurement (for mass fraction measurements greater than 0.001 g/g) was in the range of 2 % to 5 % based on a propagation of uncertainty analysis.

23

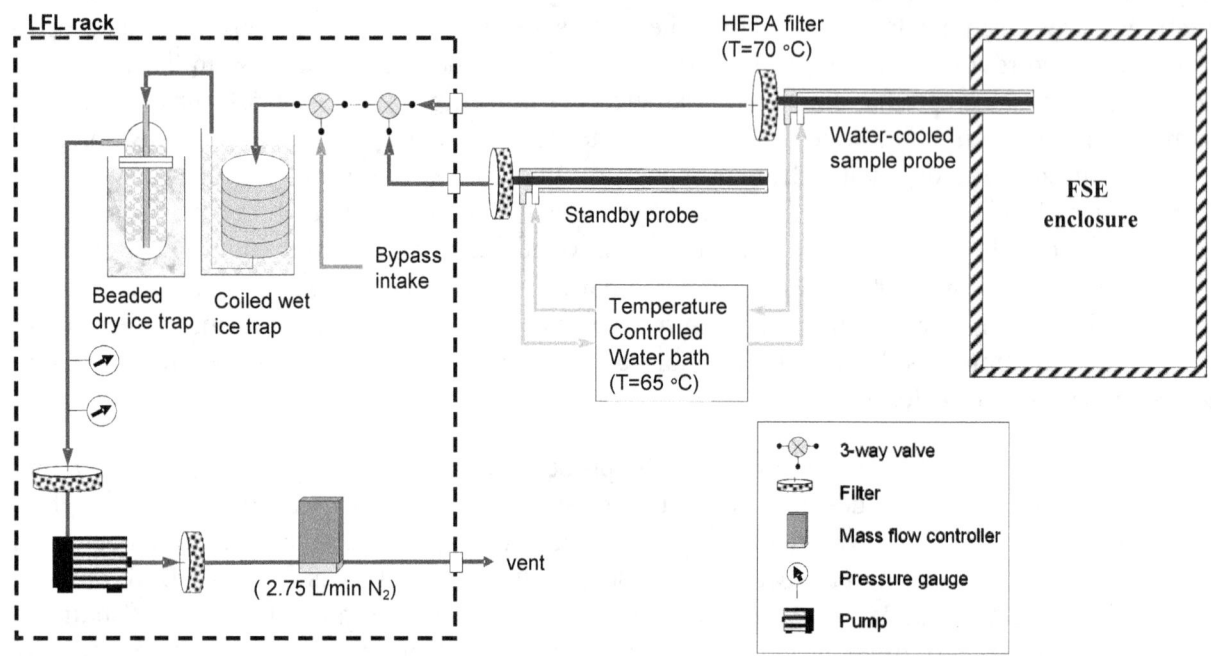

Figure 2.7: Schematic drawing of gravimetric soot sampling system

2.2.4 Bare Bead Thermocouples

Bare bead thermocouples were used to measure temperature. When using bare bead thermocouples it is important to discuss the effect of radiative losses (or gains) on the value of this measurement. Unlike aspirated thermocouples, which are specifically designed to limit radiative losses from the measurement [32], bare bead thermocouples are subject to radiative losses. This occurs, for example, when an optically thin flame (e.g. premixed methane) is being measured by a thermocouple and the surrounding ambient conditions are much colder than the flame (e.g. 2000 K flame in a 300 K room). In some of our tests, the thermocouple environment is optically thick due to heavy soot loading, and the thermocouple does not radiatively 'see' a cool surface, such as the vent of the FSE. However, there are some cases in which thermocouples may read very high temperatures in an optically thin environment with significant radiative exchange through the door with the ambient room conditions. Taking into account this component of uncertainty in addition to random variations and the inherent accuracy of the thermocouple, a combined expanded uncertainty of -20 % to +6 % with a coverage factor of 2 is reported in Table 2.2 [32].

2.2.5 Heat Flux Gauges

Total heat flux was measured at six locations during each experiment. The heat flux gauges were 6.4 mm diameter Schmidt-Boelter type, water cooled gauges with embedded type-K thermocouples. The view angle of these gauges was large, 130°. The particular model information is contained in Appendix B. The nominal range for the gauges was 150 kW/m^2. Schmidt-Boelter gauges measure a temperature difference across a thin insulating material using a thermopile to generate a voltage from the small temperature difference. These gauges typically generate voltages much less than 100 mV, even for heat fluxes near their maximum range.

Floor heat flux gauges were located in three places, just outside the doorway on the centerline of the floor ($y = -20$ cm) and straddling the burner at $y = 90$ cm and $y = 270$ cm. Each was inserted in the floor flush with the upper surface and facing vertically upward. The ceiling heat flux gauges were located at a height of 233 cm and $y = 90$ cm, $y = 180$ cm, and $y = 270$ cm along the centerline. The exact location coordinates for the gauges are listed in Table 2.1. The condition of the installed gauges was checked periodically. If significant soot accumulated on a gauge, it was brushed off after the experiment. If a gauge was no longer flush with the surface of the floor, a note was made, but there was no attempt to move the gauge since the gauges were very difficult to access.

The heat fluxes occasionally reached beyond the stated range of the gauges. According to the manufacturer, the calibrations remain linear and valid beyond the stated range as long as the materials do not degrade and change the sensitivity of the gauge. Previously the calibration of the gauges had been checked after experiencing these large heat fluxes [5]. Each gauge's responsivity was found to remain within 3 % of the factory calibration.

The main sources of uncertainty related to the total heat flux measurements are: the calibration, soot and dust deposition, and shifting of the gauge surface below the floor. These sources will be described and the combined uncertainty estimated for the reported measurements. A model of uncertainty for heat flux gauge measurements in fire environments can be found in the study by Bryant et al. [33].

The total heat flux gauge calibration from the manufacturer was used to convert mV readings to kW/m^2. This calibration was performed using cooling water at 23 °C ± 3 °C. The cooling water in the Large Fire Laboratory was found to be within the same range. The manufacturer reported a ±3 % expanded uncertainty in the responsivity (the slope in $kW/m^2/mV$) [33]. Calibrations at the NIST facility have varied within the 3 % range of the nominal manufacturer's calibration. A recent round-robin study of heat flux gauge calibration consistency [34] sent the same heat flux gauges to multiple laboratories around the world and found that while several calibrations fell within the 3 % range, when outlier data were included, then the uncertainty rose to around 8 %. For this project, an uncertainty of ±6 % for gauge calibration was chosen as fairly conservative since the NIST calibrations were within the 3 % range of the factory calibration and the range measured in the round-robin study.

Heat flux uncertainty due to soot deposition is difficult to quantify. For the tests where ethanol was burned, there was little to no contact with soot or combustion products. For the sootier fuel, heptane, at low HRRs, the lower layer remained as air with little opportunity for soot-laden gases to contact the gauges. Some soot was observed on the heat flux gauges after the experiments. For these periods of time, it was estimated that the soot coating on the gauge would add an additional uncertainty of ±10 % due to variations in surface emissivity, and soot insulating of the gauge.

The physical shifting of the gauge surface below the floor could impact the measurement if the solid angle viewed by the gauge was diminished. Since the gauge is not sensitive either in calibration or application to radiation at angles close to the plane of the gauge surface due to reflection, and the radiation approaching from the lowest angles is generally from the coolest regions of the enclosure, the gauge would have to be below the surface of the floor by a few millimeters or more to have a significant impact on its measurement. Neither gauge was ever observed to be shifted by that amount in the course of testing.

2.3   Moving Probe Sampling Locations

The gas species and temperatures were measured at various locations in the enclosure. Three movable probes were positioned along the centerline of the room, $x = 120$ cm. The probes were located at y = 0.85 m, y = 2.25 m, and y = 3.30 m. The measurement probes were attached to an exterior motor driven, linear positioning stage, capable of moving the probes along the z axis. The experiment was then repeated with the probes placed near one of the side walls of the room, $x = 10$ cm. The initial placement of the linear positioning stage resulted in the greatest uncertainty in the measurement locations. The initial positioning in the enclosure was measured by hand, resulting in a combined uncertainty at any position of ± 50 mm in any direction. The positioning stage has a short term repeatability of 0.0025 mm and straight line accuracy of 0.025 mm per 25 cm in the z direction [35]. The initial position was checked at the beginning of each experiment to ensure accuracy. An FDS simulation was utilized to determine the locations of the probes in the room which would most likely capture locations that represented the greatest extremes of the carbon monoxide production and some locations of carbon monoxide production representative of the room as a whole. Figure 2.8 presents the three dimensional contours of carbon monoxide produced by a simulated heptane fueled pool fire in the ISO 9705 room with a 10 cm doorway. The simulation in Figure 2.8 was used as guidance in selecting which x and y positions were used for measurement placement. The measurement locations used for ethanol and heptane tests are listed in Table 2.1. The vertical placement of points varied slightly from test to test, but the points were generally similar and the exact location of each point is included in the data files. Because the FDS simulations showed symmetric behavior near the left and right side walls (the $x = 0$ and $x = 2.4$ m walls), symmetry is assumed in the experiments. So the measurements taken at $x = 10$ cm in the experiments were copied and plotted at $x = 230$ cm. The measurement locations for heptane and ethanol, including the duplicated data points, are illustrated graphically in Figure 2.9 and Figure 2.10.

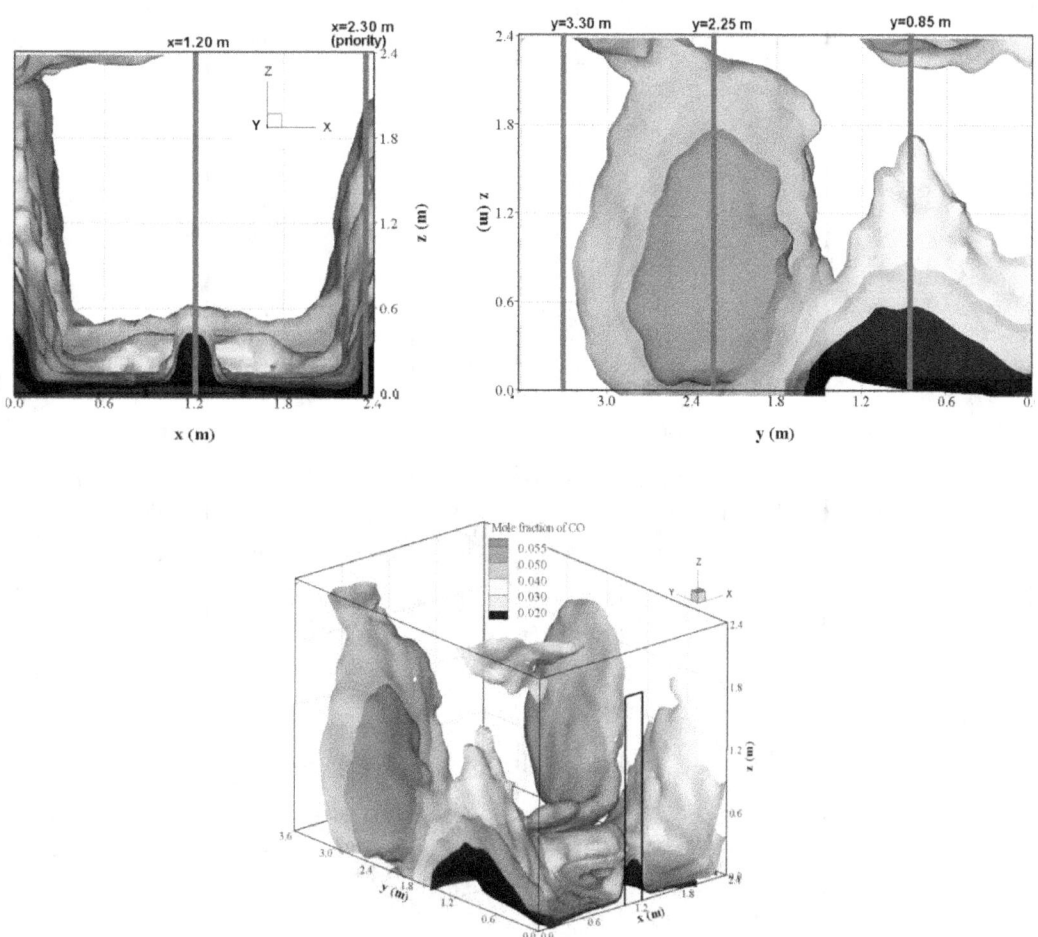

Figure 2.8: Front, left, and three dimensional view of a FDS simulation of carbon monoxide concentrations within the room used as a basis for determining optimal probe placement. The simulated case is of a heptane pool fire with a HRR of 1000 kW with a 10 cm doorway opening.

Table 2.1: Moving thermocouple and gas species analyzer locations

Heptane						Ethanol					
x (cm)	y (cm)	z (cm)	x (cm)	y (cm)	z (cm)	x (cm)	y (cm)	z (cm)	x (cm)	y (cm)	z (cm)
10	85	8	10	85	83	10	85	8	10	225	73
120	85	8	120	85	83	120	85	8	120	225	73
10	225	8	10	225	83	10	225	8	10	330	73
120	225	8	120	225	83	120	225	8	120	330	73
10	330	8	10	330	83	10	330	8	10	85	88
120	330	8	120	330	83	120	330	8	120	85	83
10	85	13	10	85	93	10	85	13	10	225	88
120	85	13	120	85	93	120	85	13	120	225	83
10	225	13	10	225	93	10	225	13	10	330	88
120	225	13	120	225	93	120	225	13	120	330	83
10	330	13	10	330	93	10	330	13	10	85	103
120	330	13	120	330	93	120	330	13	120	85	98
10	85	18	10	85	103	10	85	18	10	225	103
120	85	18	120	85	103	120	85	18	120	225	98
10	225	18	10	225	103	10	225	18	10	330	103
120	225	18	120	225	103	120	225	18	120	330	98
10	330	18	10	330	103	10	330	18	10	85	113
120	330	18	120	330	103	120	330	18	120	85	118
10	85	23	10	85	118	10	85	23	10	225	113
120	85	23	120	85	123	120	85	23	120	225	118
10	225	23	10	225	118	10	225	23	10	330	113
120	225	23	120	225	123	120	225	23	120	330	118
10	330	23	10	330	118	10	330	23	10	85	123
120	330	23	120	330	123	120	330	23	120	85	128
10	85	28	10	85	133	10	85	28	10	225	123
120	85	28	120	85	133	120	85	28	120	225	128
10	225	28	10	225	133	10	225	28	10	330	123
120	225	28	120	225	133	120	225	28	120	330	128
10	330	28	10	330	133	10	330	28	10	85	143
120	330	28	120	330	133	120	330	28	120	85	143
10	85	33	10	85	148	10	85	38	10	225	143
120	85	33	120	85	148	120	85	38	120	225	143
10	225	33	10	225	148	10	225	38	10	330	143
120	225	33	120	225	148	120	225	38	120	330	143
10	330	33	10	330	148	10	330	38	10	85	163
120	330	33	120	330	148	120	330	38	120	85	168
10	85	43	10	85	163	10	85	48	10	225	163
120	85	48	120	85	163	120	85	48	120	225	168
10	225	43	10	225	163	10	225	48	10	330	163
120	225	48	120	225	163	120	225	48	120	330	168
10	330	43	10	330	163	10	330	48	10	85	183
120	330	48	120	330	163	120	330	48	120	85	183
10	85	53	10	85	178	10	85	58	10	225	183
120	85	53	120	85	178	120	85	58	120	225	183
10	225	53	10	225	178	10	225	58	10	330	183
120	225	53	120	225	178	120	225	58	120	330	183
10	330	53	10	330	178	10	330	58	10	85	200
120	330	53	120	330	178	120	330	58	120	85	200
10	85	63	10	85	193	10	85	73	10	225	200
120	85	63	120	85	193	120	85	73	120	225	200
10	225	63	10	225	193						
120	225	63	120	225	193						
10	330	63	10	330	193						
120	330	63	120	330	193						
10	85	73	10	85	200						
120	85	73	120	85	200						
10	225	73	10	225	200						
120	225	73	120	225	200						
10	330	73	10	330	200						
120	330	73	120	330	200						

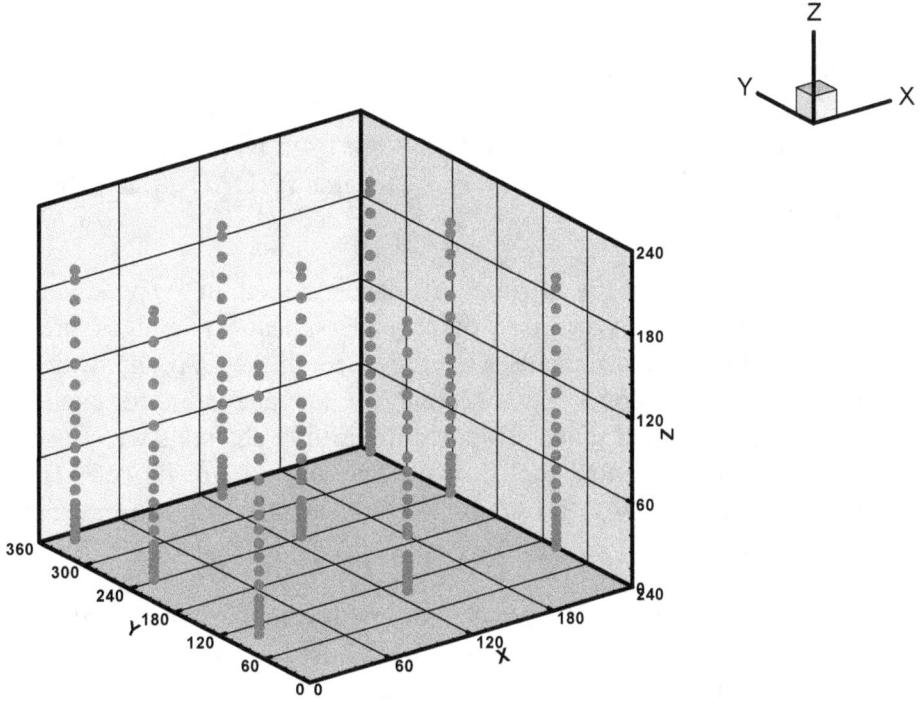

Figure 2.9: Plot of moving thermocouple and gas species analyzer measurement locations for heptane

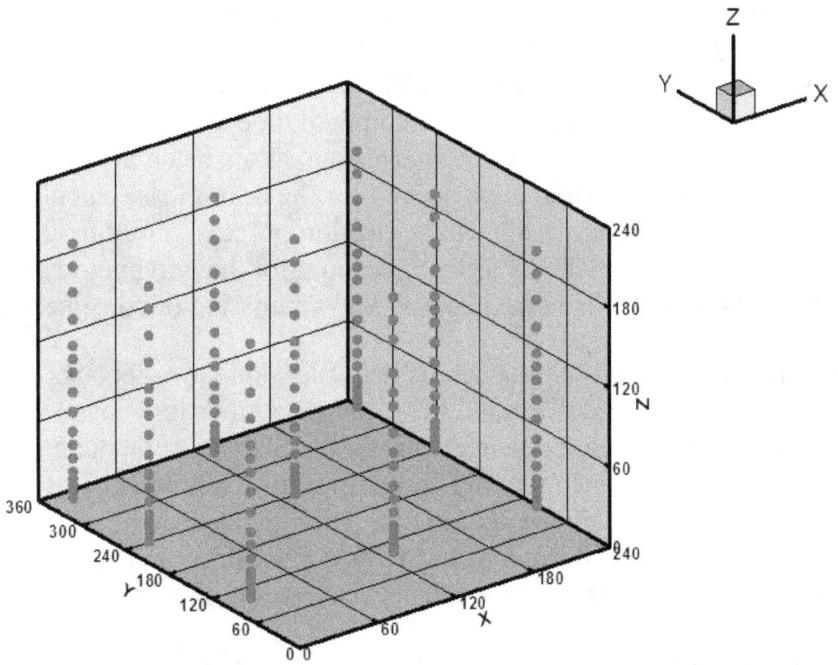

Figure 2.10: Plot of moving thermocouple and gas species analyzer measurement locations for ethanol

2.4 Data Acquisition

Two data acquisition (DAQ) systems were used in this series of experiments. One DAQ system was dedicated to fuel flows, oxygen depletion calorimetry, and the constituent measurements required to calculate the heat release rate. The other DAQ system was used to record signals from all other measurements [5]. Each DAQ system used National Instruments hardware and was controlled with LabVIEW software. The calorimetry DAQ system has been previously described in detail [6,31].

For this series of experiments, the channel list is contained in Appendix A. The types of measurements included: gas analyzers, dew point readers, heat flux gauges, pressure transducers, and thermocouples. These measurements were recorded on the DAQ hardware as voltages with 200 samples recorded every second. Each second, the average value for each channel was then converted to meaningful physical units. Two event marking channels were used to note the time of important events such as ignition, fuel flow change, or extinguishment. These event marker channels, which were in both DAQ programs, were especially useful in synchronization of the two data sets.

The data acquisition hardware had 24 bit precision, with stated accuracies of the data acquisition board and multiplexing module equal to 0.014 % and 0.015 % of the reading. These uncertainties were orders of magnitude lower than those from other sources in all of the measurements reported here.

2.5 Data Post-Processing

A Matlab script file was created for post-processing all data files generated during the test series. This program was used to make corrections to the data (including delay times and post processing not possible in real time), generate plots, and save results to ASCII text files for archival purposes. The program was also used to compute time averaged values and uncertainties for examining trends in the data. An input file was used to allow batch processing of the raw data files. The input file contained the parameters needed for the heat release calculation (this file was also read by the DAQ program during the data collection process). Additional parameters were added to the end of the standard HRR input file to account for the gravimetric soot measurements and to record the time windows when channels had known missing or corrupted data.

The first step in data reduction was to inspect the data files and lab notebooks for erroneous data resulting from open channels, loss of sample flow, or some other instrument or data acquisition malfunction. Because data were collected on two separate computers, the series were synchronized to a common reference time. The ignition time was marked using a virtual event channel on each computer and defined as time zero for the reduced data. The gas analyzer measurements from inside the FSE and exhaust hood measurements were shifted in time to account for the sample flow transfer (delay) time as discussed in section 2.2.2.

Corrections to the heat release rate measurements were applied to account for the exhaust flow calibration factor and drift in the oxygen analyzer.

Since the gases sampled from the FSE were dried before entering the detectors, an estimate of the water removed was made to correct the measurements to the *in situ* wet volume fraction. In this report, the wet volume fraction of gases is only used to determine the mixture fraction values, see section 4.1. Other gas species measurements are reported on a dry basis.

2.6 Uncertainty Summary

There are different components of uncertainty in the temperatures, total heat flux, soot mass fraction, chemical species, and heat release rate reported here. Uncertainties are grouped into two categories according to the method used to estimate them. Type A uncertainties are evaluated by statistical methods and type B are evaluated by other means [36]. Type B analysis of systematic uncertainties involves estimating the upper (+a) and lower (-a) limits for the quantity in question such that the probability that the value would be in the interval (±a) is 95 percent. After estimating uncertainties by either Type A or B analysis, the uncertainties are combined in quadrature to yield the combined standard uncertainty. Multiplying the combined standard uncertainty by a coverage factor of two results in the total expanded uncertainty that corresponds to a 95 percent confidence interval (2σ).

Components of uncertainty were tabulated in Table 2.2. The largest uncertainty component of moving probe location came from the initial placement of the linear positioning stage, which was measured by hand. The stage was far more precise. Some of these components, such as the zero calibration elements, are derived from instrument specifications. Other components, such as radiation loss and thermophoretic soot deposition included past experience with these measurements. The uncertainty in the temperature measurements, depending on measurement's location, did include radiative cooling, which likely resulted in a measured temperature lower than the actual gas temperature. Part of the uncertainty was attributed to the accuracy of the mass scale used to weigh the soot filters, see section 2.2.3, and part of the uncertainty was due to uncertainties in the flow rate which was measured after the gases had been cooled and was therefore highly dependent on the temperature at the entrance to the soot probe. Additional uncertainty was introduced by the water correction discussed in section 2.5. Uncertainties in the heat release rate measurement can be traced to variations in the hood duct flow profile, soot and total hydrocarbons which are not accounted for, and small instrument uncertainties. The general function of the heat release rate measurement is discussed in section 2.2.1 and the uncertainty of the measurement in this hood has been well documented [31]. The associated uncertainty in the ideal heat release rate, used to calculate combustion efficiency is related primarily to the purity of the fuel and the uncertainty in the fuel flow rate measurement device, both of which are small. The gas analyzers, $CO/CO_2/O_2/THC$, used precision mixed zero and span gases and have a small uncertainty reported by the manufacturer. For these devices the random and mixing/averaging due to long sample lines are a larger source of uncertainty. The heat flux gauges used here were generally very precise devices and despite being used beyond their calibrated range have been shown to be quite linear and repeatable, cf. section 2.2.5. The larger sources of uncertainty arose as a result of the cooling water being unable to remove heat from the heat flux gauge fast enough and because of thermophoretic soot deposition on the heat flux gauge surface. The uncertainties are reported as a range in Table 2.2 and are represented by error bars in the associated figures.

Table 2.2: Summary of uncertainty of measurements

	Component Standard Uncertainty	Combined Standard Uncertainty	Total Expanded Uncertainty
Moving Probe Location			(x, y, & z direction) ± 50 mm

Temperature
   Calibration
   Radiative Cooling
   Random
	
	± 1 %
	-10 % to 0 %
	± 3 %
		-10 % to +3%	-20 % to +6 %

Heat Flux
   Soot Deposition
   Cooling Water Temp
   Calibration
   Random
	
	-5 % to 0 %
	± 5 %
	± 1 %
	± 3 %
		-8 % to +6 %	-16 % to +12 %

Gas Analyzers
   Zero and Span Gas
   Equiptment Uncertainty
   Mixing and Averaging
   Random
	
	± 1 %
	± 1 %
	± 5 %
	± 3 %
		± 6 %	± 12 %

Soot Mass Fraction
   Mass Measurement
   Volume Flow Rate
   Random
	
	± 1 %
	± 1 %
	± 3 %
		± 3 %	± 5 %

Measured Heat Release Rate
   Exhaust Flow Rate
   Soot and THC
   Instruments Uncertainty
   Random
	
	± 5 %
	± 3 %
	± 1 %
	± 3 %
		± 7 %	± 14 %

Ideal Heat Release Rate
   Fuel Purity
   Equiptment Uncertainty
   Random
	
	± 1 %
	± 1 %
	± 3 %
		± 3 %	± 6 %

3 RESULTS

3.1 List of Experiment Conditions

Nine experiments were conducted using sprayed ethanol and heptane. The complete list of experiments is included in Table 3.1.

Table 3.1: List of experiment conditions

Experiment ID	Date	Fuel	Nozzle BETTE Model No.	Doorway Width (cm)	Lateral Probe Location (cm)	Duration (min)
ISOHept38	4/14/2009	Heptane	1-1/2	10	120	64
ISOHept39	4/15/2009	Heptane	1-1/2	10	120	57
ISOEth40	4/15/2009	Ethanol	1-1/2	10	120	34
ISOEth42	4/16/2009	Ethanol	1-1/2	10	120	74
ISOHept45	4/20/2009	Heptane	1-1/2	10	10	40
ISOHept46	4/21/2009	Heptane	1-1/2	10	10	75
ISOEth47	4/21/2009	Ethanol	1-1/2	10	10	66
ISOEth48	4/22/2009	Ethanol	1-1/2	10	10	69
ISOHept51	4/23/2009	Heptane	1-1/2	10	10	83

3.2 Heat Release Rate

As the fire became under-ventilated, burning took place outside of the enclosure. The HRR measurement represents the total burning inside and outside of the enclosure. Table 3.2 shows a description of the measurement labels used in the table column headings and figure legends to describe heat release rate in this section. These labels are identical to the column headings in the reduced data files.

The HRR was measured using oxygen calorimetry, while the ideal HRR (IHRR) was determined from the fuel delivery rate to the burner. For the purpose of creating repeatability, the IHRRs were ramped quickly to a specified IHRR. They were then maintained at a steady state by continuously pumping fuel through the spray nozzle into the pan in the center of the room. In all of the experiments, a 0.7 m × 0.7 m square spray burner was placed in the center of the room and a doorway width of 10 cm was utilized. In most of the experiments the IHRR was held constant at 1000 kW. Figure 3.1 displays the ideal and measured HRR for a representative heptane burn, ISOHept39. As expected, the measured HRRs of the experiments are generally significantly less than the ideal HRR as can be seen in Figure 3.1. For heptane, the measured HRR ranged between 600 kW and 900 kW.

Using the same configuration and HRR as the heptane experiments, four ethanol experiments were also conducted. The HRR for a representative ethanol burn, ISOEth48, can be found in Figure 3.2. The measured and ideal values for ethanol are closer to one another than they are for the heptane values. This is due to ethanol being a cleaner burning fuel than heptane and thus having a greater combustion efficiency. The measured HRR values range from 750 kW to 1050 kW.

33

A comparison of ideal and measured heat release rates for heptanes and ethanol fires is presented in Figure 3.3. The dashed line indicates a combustion efficiency of 100% of the fuel in each case. The ideal burning rate was determined from the liquid fuel flow rate and the known heat of combustion. The combustion efficiency of ethanol is generally higher than heptane.

Table 3.3 presents the mean values for the measured and ideal steady-state heat release rates for each experiment conducted.

Table 3.2: Description of calorimetry measurement labels

Measurement Label	Description
HRR	Heat Release Rate from Calorimeter, kW
IHRR	Ideal Heat Release Rate from Burner (spray), kW

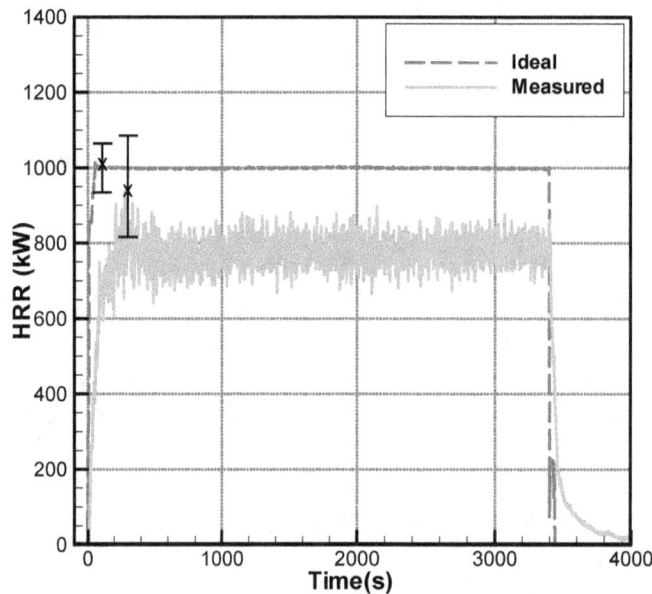

Figure 3.1: Heat release rate for test ISOHept39 (Heptane) comparing the ideal heat release rate estimated from the mass loss rate measurement assuming complete combustion (red line) and the heat release rate measured by oxygen loss calorimetry (green line)

34

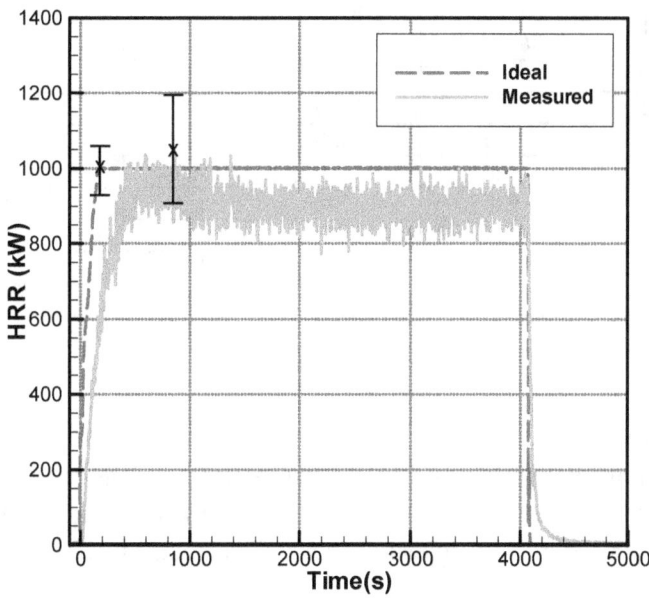

Figure 3.2: Heat release rate for experiment ISOEth48 (Ethanol) comparing the ideal heat release rate estimated from the mass loss rate measurement assuming complete combustion (red line) and the heat release rate measured by oxygen loss calorimetry (green line)

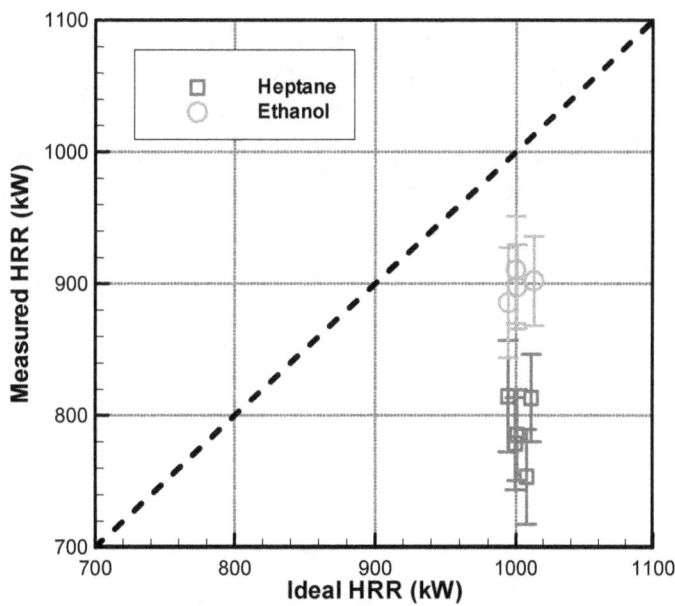

Figure 3.3: Steady state heat release results. The dashed line indicates ideal or complete burning.

Table 3.3: Summary of averaged steady state results of HRR and exhaust stack species measurements. U indicates the standard deviation in each steady state measurement.

Experiment ID	Fuel	Steady State Window		HRR (kW)		IHRR (kW)	
		Start (s)	*Stop (s)*	*Mean*	*U*	*Mean*	*U*
ISOHept38	Heptane	900	3300	754	36	1008	1
ISOHept39	Heptane	900	3300	779	35	1000	1
ISOEth40	Ethanol	600	1800	886	42	994	1
ISOEth42	Ethanol	1500	3900	902	34	1013	1
ISOHept45	Heptane	600	2100	815	42	994	5
ISOHept46	Heptane	1200	3600	813	33	1011	3
ISOEth47	Ethanol	1200	3600	911	41	1000	1
ISOEth48	Ethanol	1200	3600	898	32	1001	1
ISOHept51	Heptane	2400	4800	785	35	1001	1

3.3 Temperatures

In this study, Type K bare-bead thermocouples were used to measure temperature. Refer to Table 2.1 for exact locations of the thermocouples. The measurement labels used in the table column heading and figure legends in this section are described in Table 3.4.

To demonstrate the reproducibility of the measurements over a number of different experiments, Figure 3.4 shows averaged temperatures measured at the front and rear thermocouple arrays as a function of the height above the floor for ISOHept38, ISOHept45, and ISOHept51, all at similar average HRRs of about 780 kW (Table 3.3). In this figure, the vertical profiles of temperature at the front location for all of the cases are nearly identical. The temperature profiles at the rear location show only minor differences between the cases except for the measurements below 0.6 m. Similarly, Figure 3.5 shows the averaged temperatures measured at the front and rear thermocouple arrays as a function of height for ISOEth40, ISOEth42, and ISOEth48, each experiment with a similar average HRR of about 900 kW (Table 3.3). The temperature profiles show nearly the same results above 0.3 m with the exception of the front thermocouple array in the ISOEth40 experiment. The values in the rear vary on average 28 °C while the front values vary 84 °C. From these figures it is clear that temperature measurements were reasonably reproducible for each of the fuel types.

Table 3.4: Description of interior gas temperature measurement labels

Measurement Label	Description
TFSamp	Moving bare bead thermocouple at front sample location (85 cm into the room, away from the door)
TCSamp	Moving bare bead thermocouple at center sample location (225 cm into the room, away from the door)
TRSamp	Moving bare bead thermocouple at rear sample location (330 cm into the room, away from the door)
TF3	Bare bead thermocouple at front location(2.5 cm above floor)
TF30	Bare bead thermocouple at front location (30 cm above floor)
TF60	Bare bead thermocouple at front location (60 cm above floor)
TF90	Bare bead thermocouple at front location (90 cm above floor)
TF105	Bare bead thermocouple at front location (105 cm above floor)
TF120	Bare bead thermocouple at front location (120 cm above floor)
TF135	Bare bead thermocouple at front location (135 cm above floor)
TF150	Bare bead thermocouple at front location (150 cm above floor)
TF180	Bare bead thermocouple at front location (180 cm above floor)
TF210	Bare bead thermocouple at front location (210 cm above floor)
TF237	Bare bead thermocouple at front location (237.5 cm above floor)
TR3	Bare bead thermocouple at rear location (2.5 cm above floor)
TR30	Bare bead thermocouple at rear location (30 cm above floor)
TR60	Bare bead thermocouple at rear location (60 cm above floor)
TR90	Bare bead thermocouple at rear location (90 cm above floor)
TR105	Bare bead thermocouple at rear location (105 cm above floor)
TR120	Bare bead thermocouple at rear location (120 cm above floor)
TR135	Bare bead thermocouple at rear location (135 cm above floor)
TR150	Bare bead thermocouple at rear location (150 cm above floor)
TR180	Bare bead thermocouple at rear location (180 cm above floor)
TR210	Bare bead thermocouple at rear location (210 cm above floor)
TR237	Bare bead thermocouple at rear location (237.5 cm above floor)

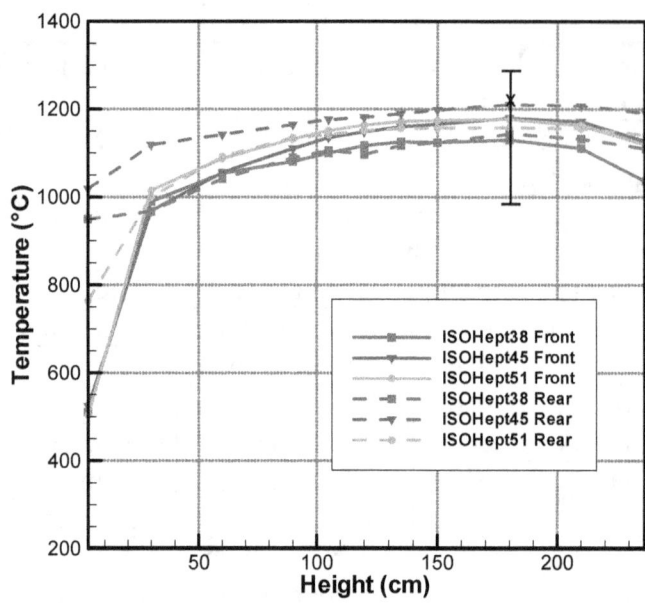

Figure 3.4: Comparisons of averaged temperature measured at front and rear thermocouple arrays for experiment ISOHept38, ISOHept45, and ISOHept51

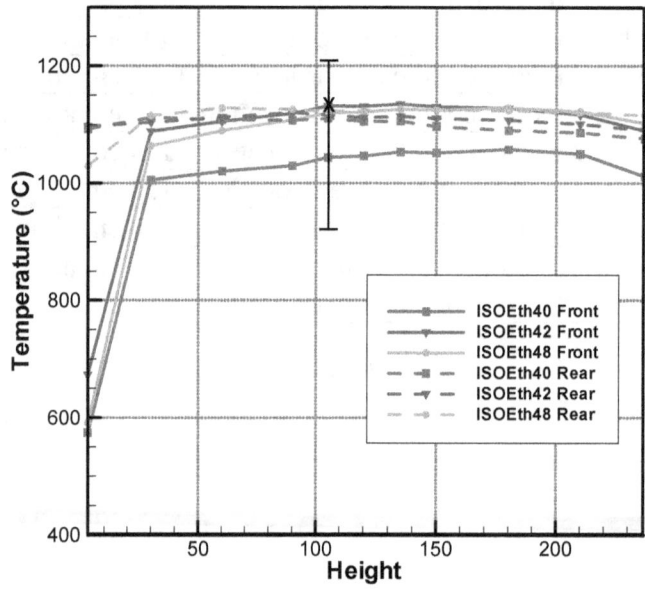

Figure 3.5: Comparisons of averaged temperature measured at front and rear thermocouple arrays for experiment ISOEth40, ISOEth42, and ISOEth48

38

During the experiments, temperature measurements were recorded at numerous locations during the steady state period of the fire in order to characterize the fire environment throughout the room. The measurements were averaged over a steady dwell time of 1 min to 2 min. The exact locations at which measurements were taken are listed in Table 2.1. Figure 2.9 and Figure 2.10 show the measurement locations. The temperature results from each location are displayed in three dimensional contour plots. The doorway in the compartment is centered at $y = 0$ cm and $x = 120$ cm in each plot. Each plot is a combination of two experiments of a particular fuel (at $x = 120$ cm and $x = 10$ cm) to provide a more complete view of the spatial distribution. Symmetry is assumed, and the $x = 10$ cm location values are duplicated on the opposite side of the room at $x = 230$ cm. The combination of data points from two experiments provides 120 points to plot for heptane and 102 points for ethanol. Linear interpolation was applied to the data to create isosurfaces. Figure 3.6 displays the temperatures taken throughout the room in ISOHept39 and ISOHept46. Figure 3.7 show the temperatures for ISOEth42 and ISOEth47. Isosurfaces at varying temperatures are marked by black lines to show the temperatures in the room. In both of the cases, the lowest temperatures were near the floor at the door opening. However, the distinct difference between the heptane and ethanol experiments is that in the heptane experiments the cooler 800 °C gases were near the floor in the area just behind the door opening. In the case of ethanol, the cooler air was just behind the entire height of the door. Also, in the heptane cases, temperatures reaching 1300 °C were measured in the rear upper corners. On the other hand in the ethanol cases, the temperature was on the order of 1100 °C and was located about 120 cm above the floor and the center of the room back to the rear wall and forward to the side walls closer to the door. The temperature data reported here show significant differences in fire environment between fuel types.

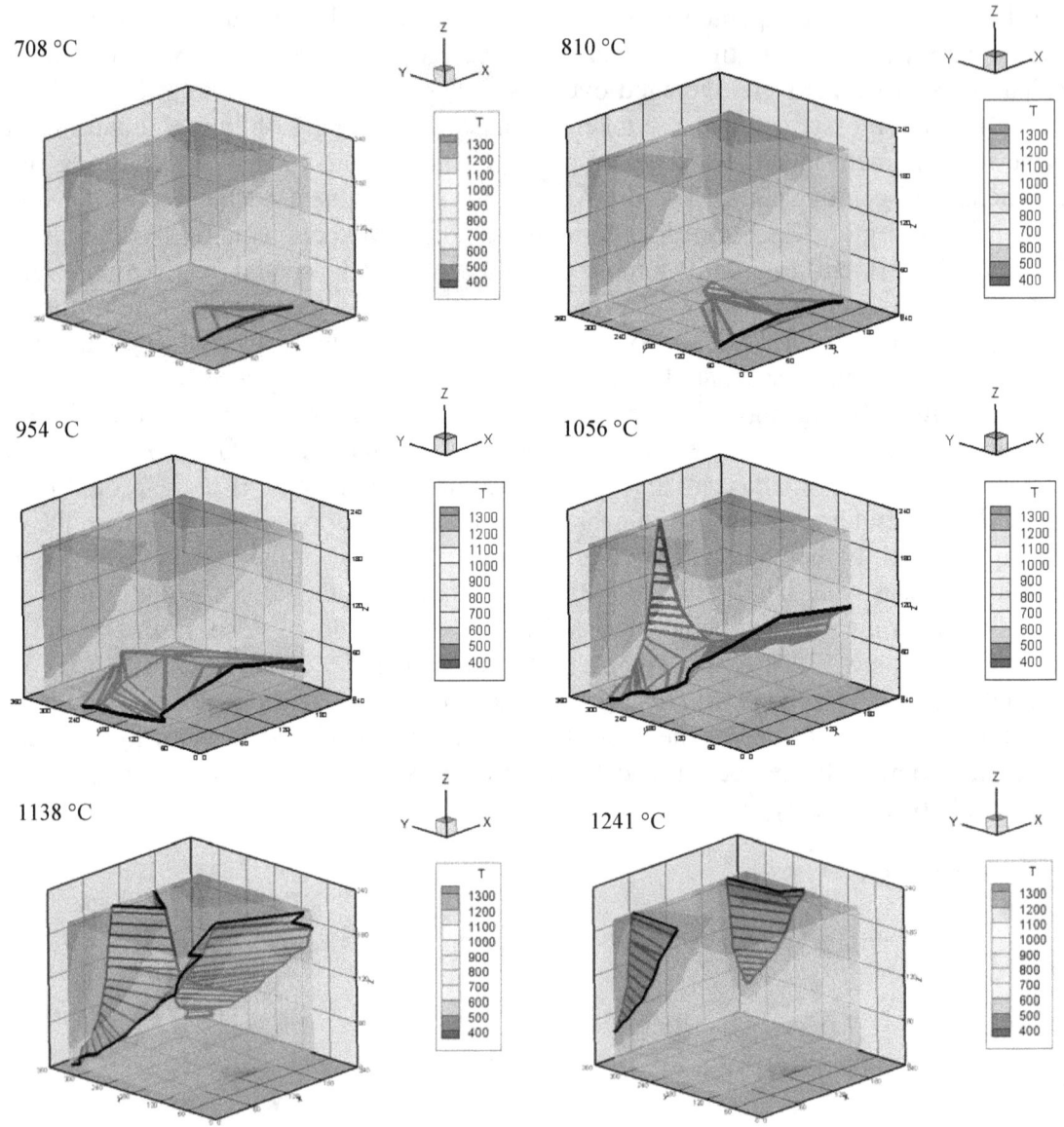

Figure 3.6: Multiple views of the interior temperature contour plot for the heptane experiments depicting isosurfaces (indicated by black lines) at various specific temperatures

40

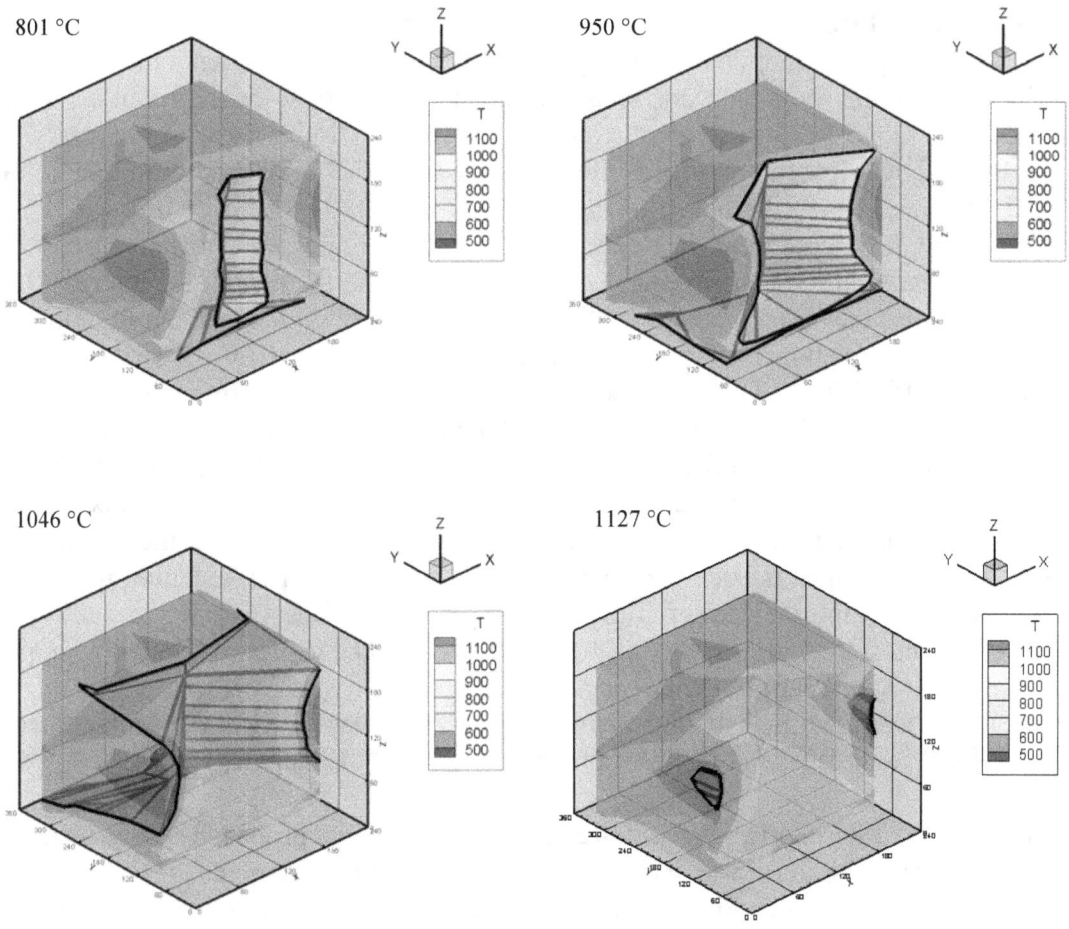

Figure 3.7: Multiple views of the interior temperature contour plot for the ethanol experiments depicting isosurfaces (indicated by black lines) at various specific temperatures

3.4 Heat Flux

Schmidt-Boelter type thermopile heat flux gauges were used to continuously measure heat fluxes at five locations in the room, and one location outside the doorway of the room as described in section 2.2.5. Three of the heptane experiments yielded usable heat flux results. In the ISOHept46 and ISOHept51 experiments the cooling water flow failed, resulting in unrealistically high heat flux readings. The heat flux results taken at the floor and ceiling for ISOHept38, ISOHept39, and ISOHept45 can be found in Figure 3.8-Figure 3.13. An explanation of the measurement labels can be found in Table 3.5. In Figure 3.8, the results for the rear floor were not plotted because they were unrealistically low, attributed to equipment malfunction. In both ISOHept39 and ISOHept45, the maximum heat flux was achieved by the rear floor heat flux gauge at about 110 kW/m^2. The observed behavior for the fluxes can be seen in the ISOHept39 and ISOHept45 figures. The rear floor gauge yielded the highest heat flux values because it was closer to the fire plume and the fire drew in fresh air near the floor of the doorway, pushing the fire towards the rear of the compartment. For that reason, the front floor heat flux measurement was lower than the rear floor gauge. The outside floor gauge yielded the lowest heat flux because it was located just outside the doorway where the air entered, meaning that the gauge's viewing angle was capturing largely ambient temperatures. The rear of the ceiling likewise experienced the most intense heat flux at the ceiling level. The center of the ceiling experienced less than the rear and the front experiences still less. ISOHept45 was unusual in that the rear ceiling heat flux was continuously increasing and was at similar levels to the center of the ceiling until about 1200 seconds where, the rear surpassed the center (Figure 3.13).

Similarly, only the ISOEth40 and ISOEth42 ethanol experiments produced usable results (Figure 3.14-Figure 3.17). The general behavior of the heat flux measurements were the same as the heptane fires with the exception that the front floor heat flux surpasses the rear floor and experiences the highest heat flux overall. In Figure 3.14 the ISOEth40 experiment resulted in a higher rear floor heat flux than the front floor heat flux for most of the plot. This was because this experiment was not allowed to run at steady-state as long as the others were. The other two corresponding ethanol plots showed similar behavior early in the plot. In each experiment, the maximum value achieved was about 140 kW/m^2.

In reports [11,37] of an earlier experimental series, 20 kg of heptane was allowed to burn in a 0.5 m^2 pan that was 10 cm deep. In that experiment the door was larger than that used here; 1/4 of the standard door width instead of 1/8. There are several noticeable differences between the heat flux results from the free burn experiment and the results found here. Because the heat release rate was fairly steady for the current experiments, the heat fluxes remained fairly constant as well. Here the rear floor heat flux was continuously higher than the front floor heat flux during the entire heptane experimental period. On the floor in the previous experiment, the front heat flux started below the rear heat flux and increased with time, eventually surpassing the rear flux about 250 s into the experiment. On the ceiling in the controlled heptane fires the rear experienced the highest heat fluxes followed by the center and then the front locations. In the previous experiment, the center position experienced the highest followed by the front and rear. Like the floor values, the front ceiling values surpass the center ceiling values at about 250 s into the experiment. The highest sustained heat flux for heptane was about 110 kW/m^2. The heptane experiment from the previous report reached a maximum heat flux of about 150 kW/m^2.

42

Table 3.5: Description of heat flux measurement labels

Measurement Label	Description
HFRFL	Heat flux gauge at rear floor location facing up, kW/m^2
HFFFL	Heat flux gauge at front floor location facing up, kW/m^2
HFOFL	Heat flux gauge at outside floor location facing up, kW/m^2
HFRCE	Heat flux gauge at rear ceiling location facing down, kW/m^2
HFCCE	Heat flux gauge at center ceiling location facing down, kW/m^2
HFFCE	Heat flux gauge at front ceiling location facing down, kW/m^2

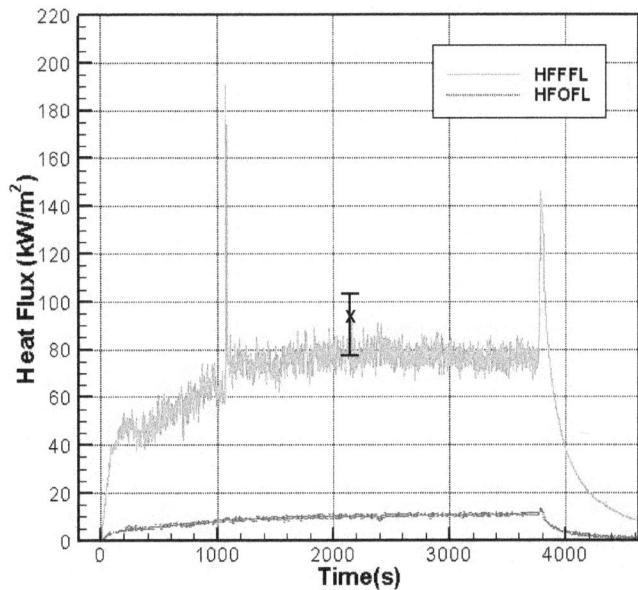

Figure 3.8: Comparison of heat flux measurements made on the floor for ISOHept38

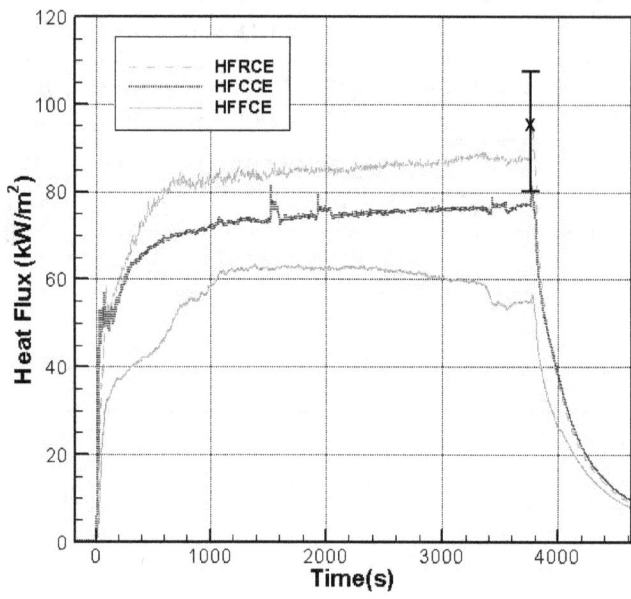

Figure 3.9: Comparison of heat flux measurements made at the ceiling for ISOHept38

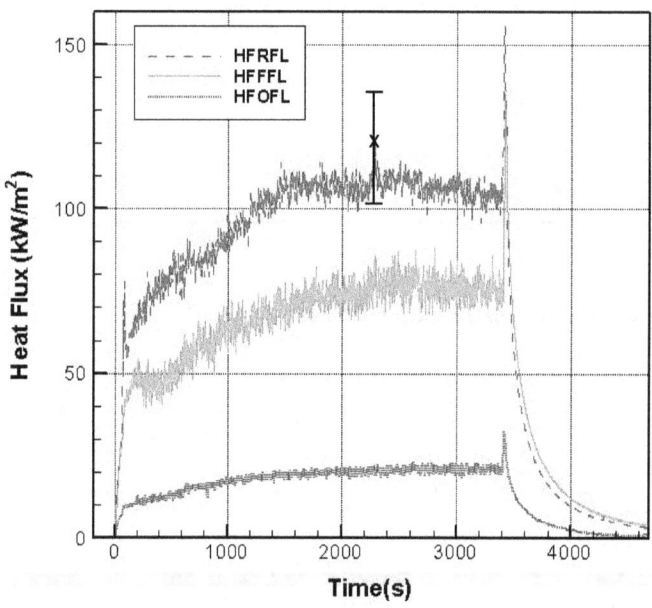

Figure 3.10: Comparison of heat flux measurements made on the floor for ISOHept39

44

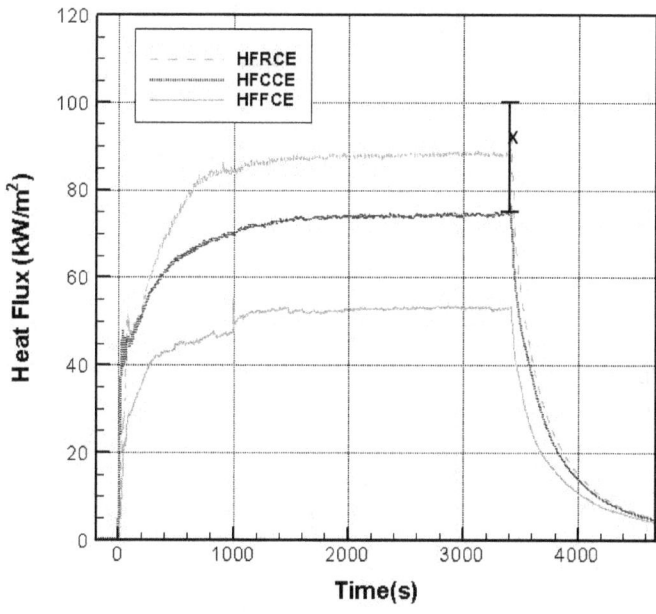

Figure 3.11: Comparison of heat flux measurements made at the ceiling for ISOHept39

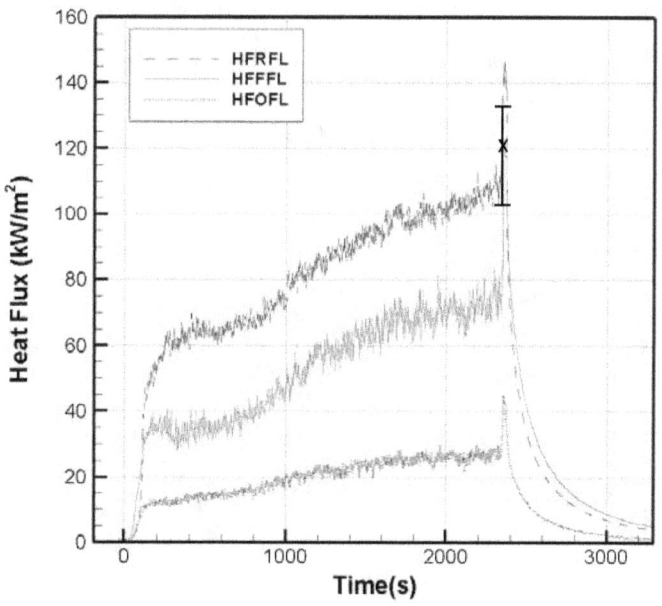

Figure 3.12: Comparison of heat flux measurements made on the floor for ISOHept45

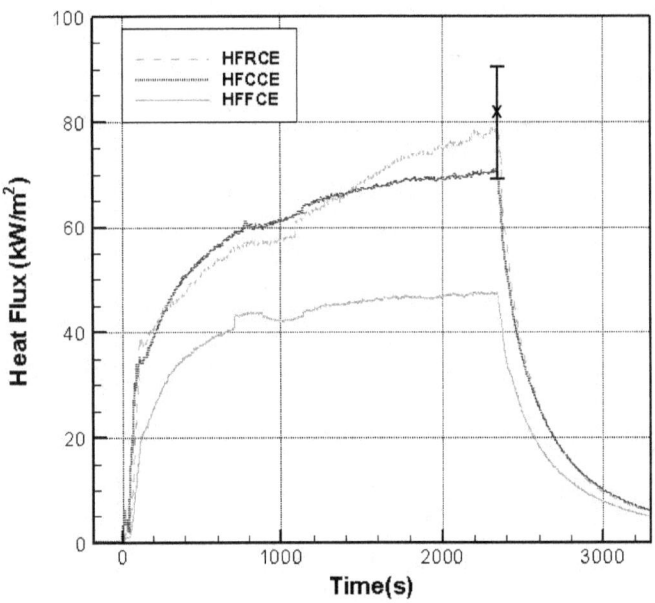

Figure 3.13: Comparison of heat flux measurements made at the ceiling for ISOHept45

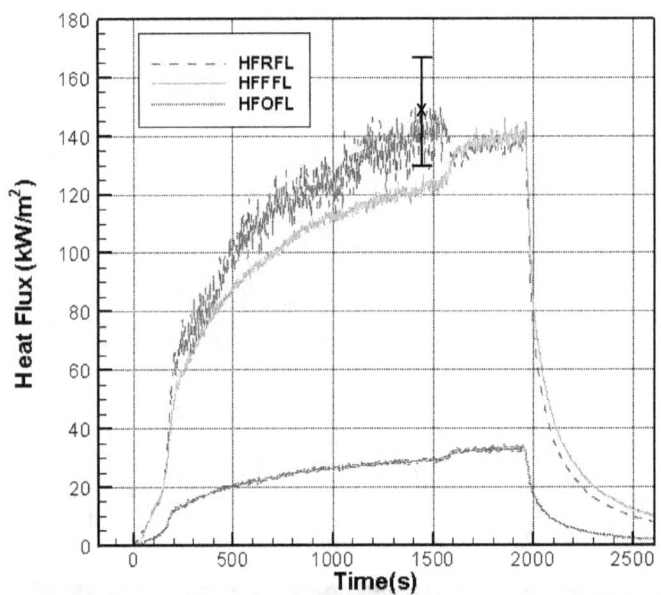

Figure 3.14: Comparison of heat flux measurements made on the floor for ISOEth40

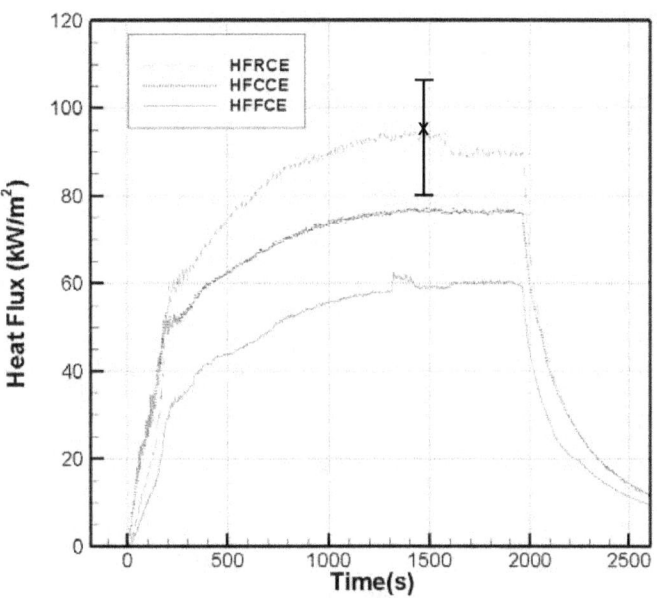

Figure 3.15: Comparison of heat flux measurements made at the ceiling for ISOEth40

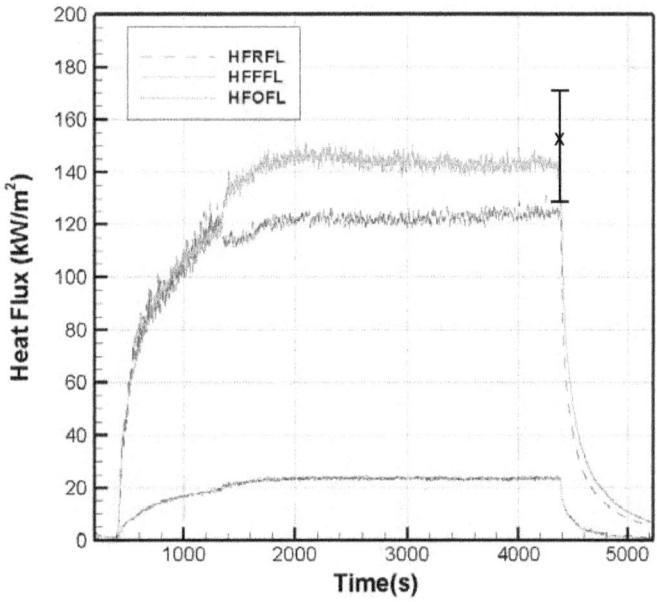

Figure 3.16: Comparison of heat flux measurements made on the floor for ISOEth42

47

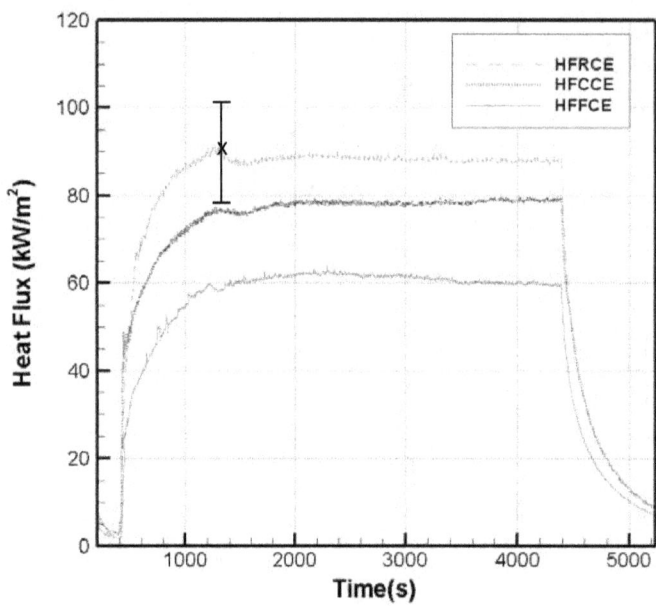

Figure 3.17: Comparison of heat flux measurements made at the ceiling for ISOEth42

3.5   Interior Compartment Gas Species

All gas species measurements are reported on a wet basis. O_2, CO, CO_2, and total hydrocarbons (THC) were monitored continuously by the gas analyzers discussed in section 2.2.2. The sampling probes for the gas analyzers were moved to multiple positions within the compartment in conjunction with the transient temperature readings to get gas species data at points throughout the room during the steady state burning periods in order to construct three dimensional plots. The times at which the measurements were taken can be disregarded because during the steady-state periods minimal variations are observed in the HRRs and temperatures. The three dimensional plots provide a glimpse into the spatial behavior of the gas species inside a compartment fire. The heptane figures in this section are a combination of the gas species data taken in ISOHept39 and ISOHept46. Similarly, the ethanol figures consist of the results from ISOEth42 and ISOEth47. These experiments were chosen to be combined based on the similarity of their global parameters and the number of measurement points they had in common. In the figures, black lines mark the isosurfaces for varying volume fractions. A list of the locations where measurements were taken can be found in Table 2.1. Three dimensional plots of the locations are shown in Figure 2.9 and Figure 2.10.

Figure 3.18 displays the O_2 volume fraction contour plot for heptane. The highest levels of oxygen were located behind the door opening at the floor and along the back half of the floor near the left and right side walls. The lowest levels of oxygen occurred from 60 cm off the floor to the ceiling behind the doorway and back around the x-axis and extending over much of the room. Figure 3.19 shows the O_2 volume fraction contour plot for ethanol. The highest oxygen levels were found around the edges of the room near the floor, and the lowest oxygen levels were in the upper half of the room. Both fuels showed a quick drop in oxygen levels by the door. At $y = 85$ cm and $x = 120$ cm the levels dropped from the volume fraction of oxygen in ambient air to nearly zero between the floor and $z = 24$ cm. There were also high levels of oxygen near the rear floor corners of the room and little to no oxygen at the front ceiling. In both cases the areas with the highest O_2 concentrations were near locations where fresh air is entering the room. The same areas in which the oxygen levels were highest, contained very little to no amounts of the products of combustion as seen in Figure 3.20-Figure 3.27.

The CO_2 volume fraction contour plot for heptane can be found in Figure 3.20. The CO_2 concentration was lowest around the rear of the burner on the floor, which was centered at $x = 180$ cm and $y = 180$ cm. The highest volume fractions were about 0.1. This area stretched from the rear ceiling corners to midlevel edges of the front wall. Figure 3.21 shows the carbon dioxide levels for the ethanol experiments. The lowest levels again were found in the vicinity of the burner. The highest concentrations were on the order of 0.1. These concentrations manifested themselves in the front half of the room on either side of the enclosure near the burning layer interface.

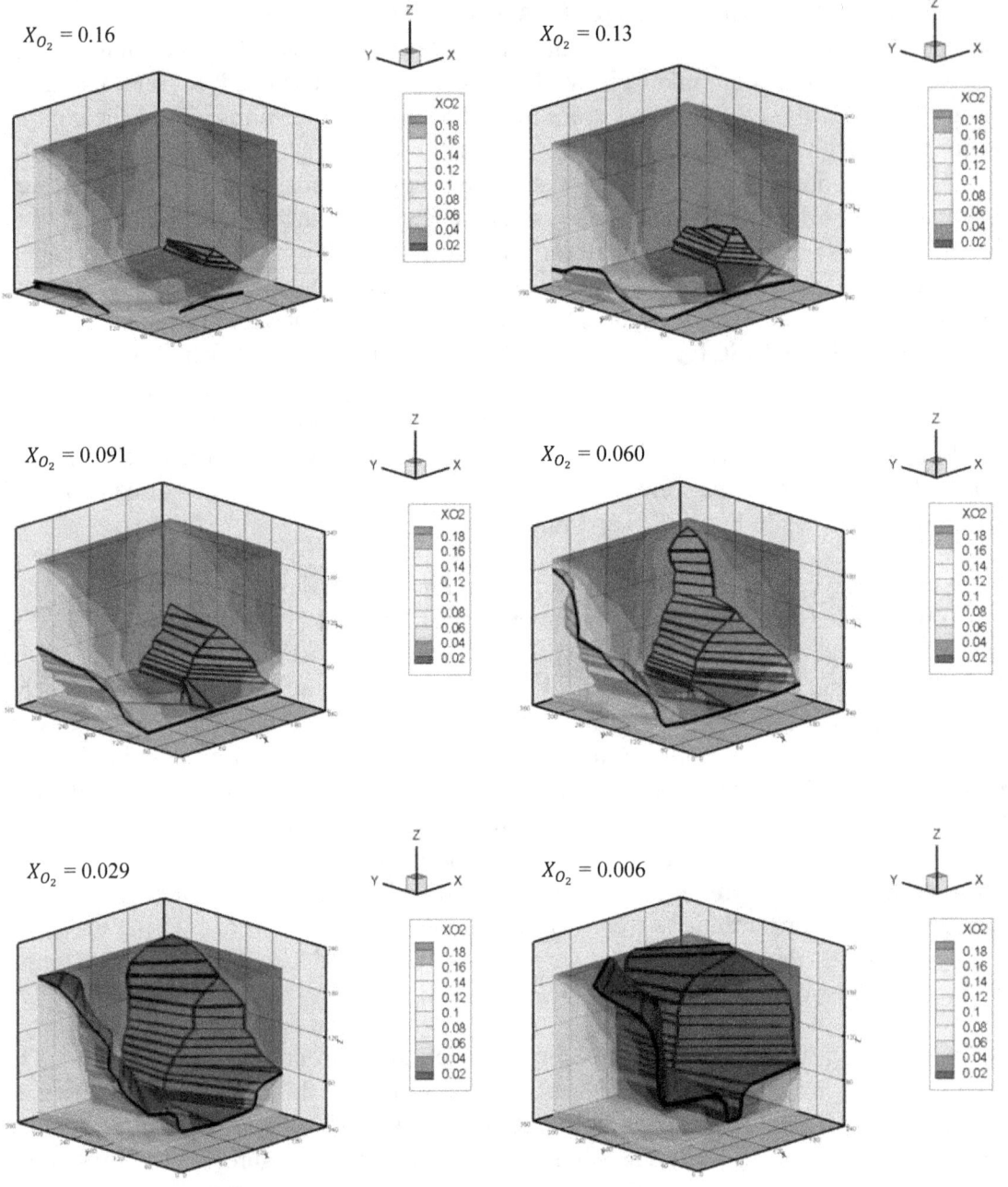

Figure 3.18: Multiple views of the interior oxygen volume fraction contour plot for the heptane experiments depicting isosurfaces (indicated by black lines) at various specific volume fractions

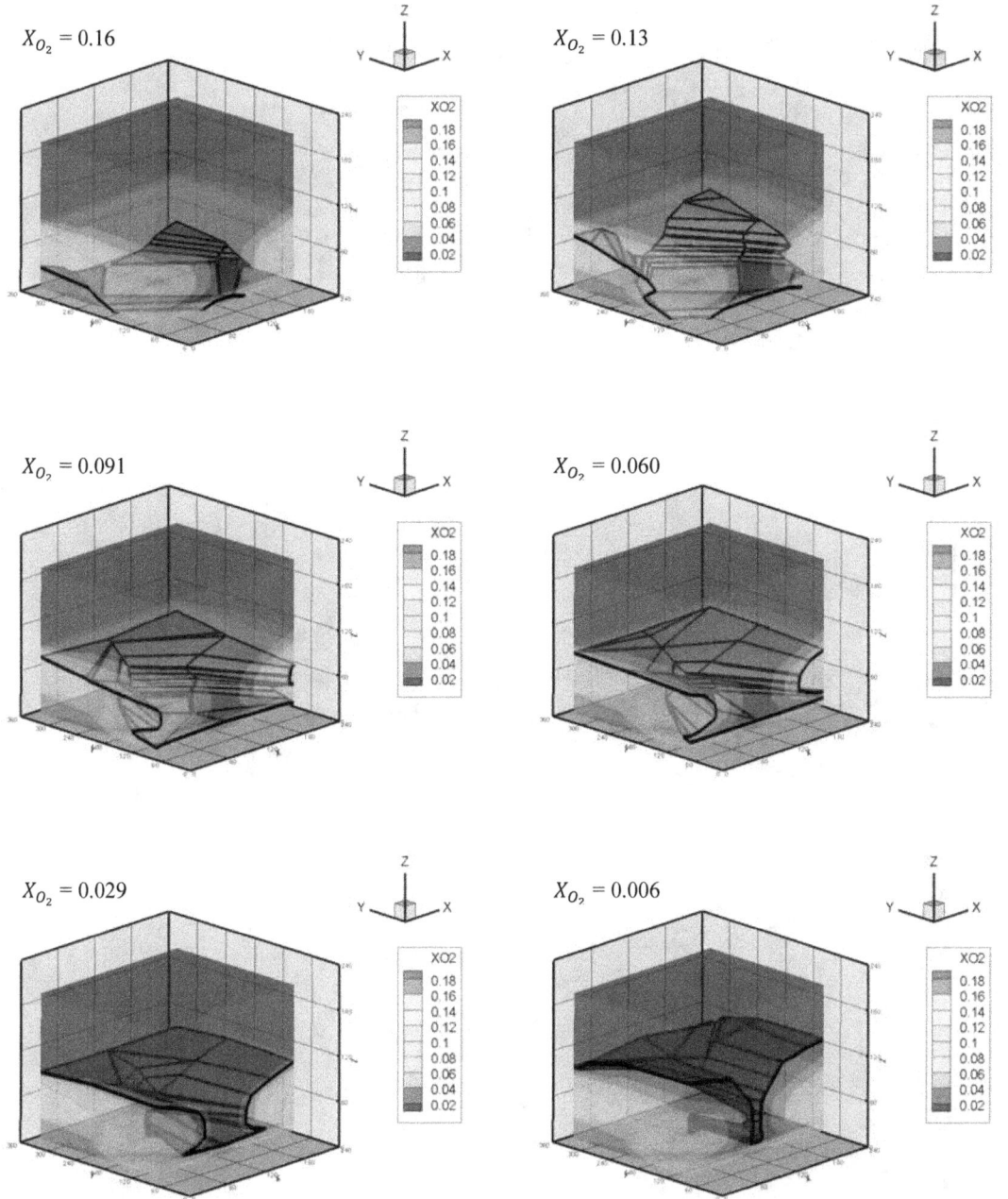

Figure 3.19: Multiple views of the interior oxygen volume fraction contour plot for the ethanol experiments depicting isosurfaces (indicated by black lines) at various specific volume fractions

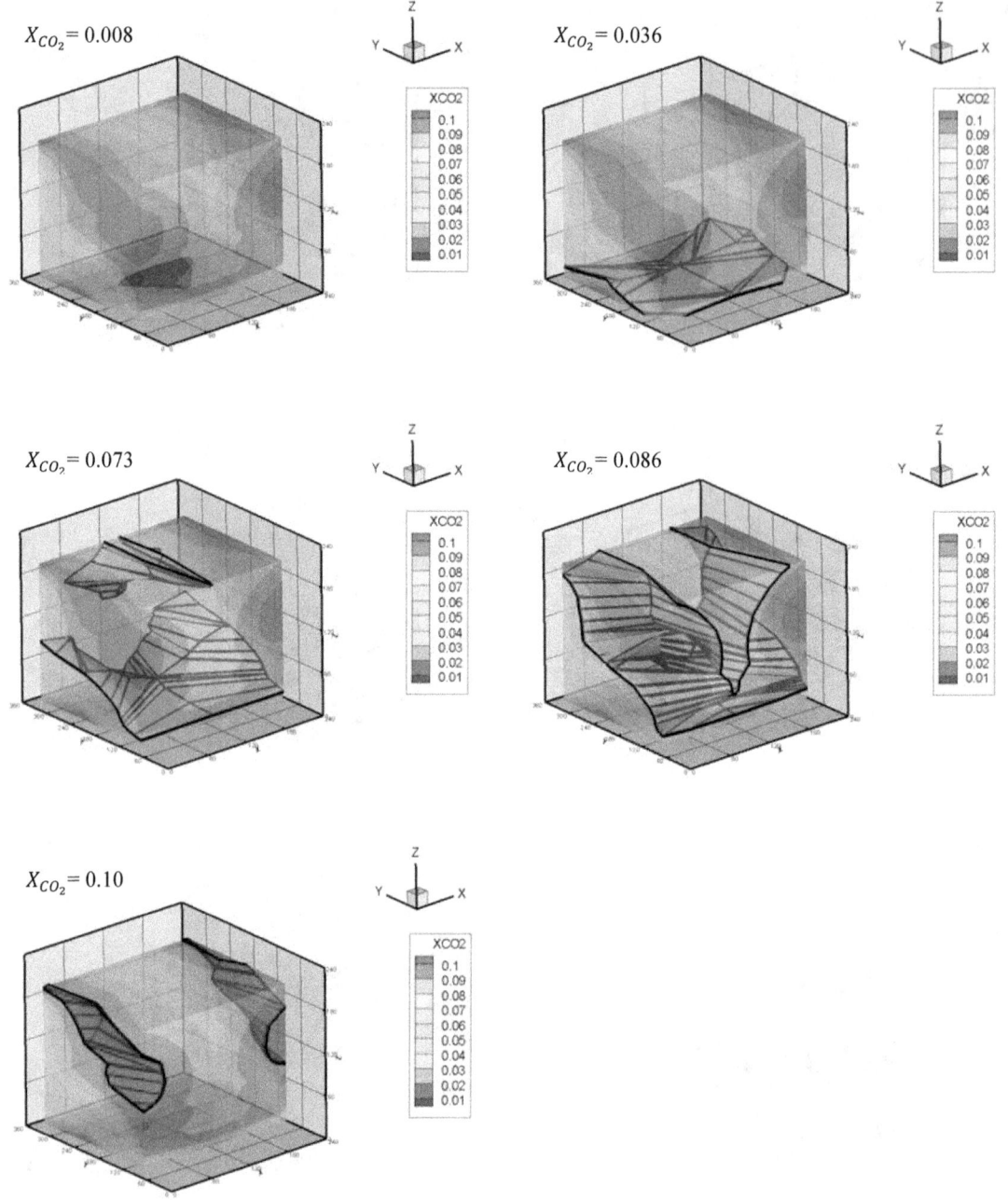

Figure 3.20: Multiple views of the interior carbon dioxide volume fraction contour plot for the heptane experiments depicting isosurfaces (indicated by black lines) at various specific volume fractions

52

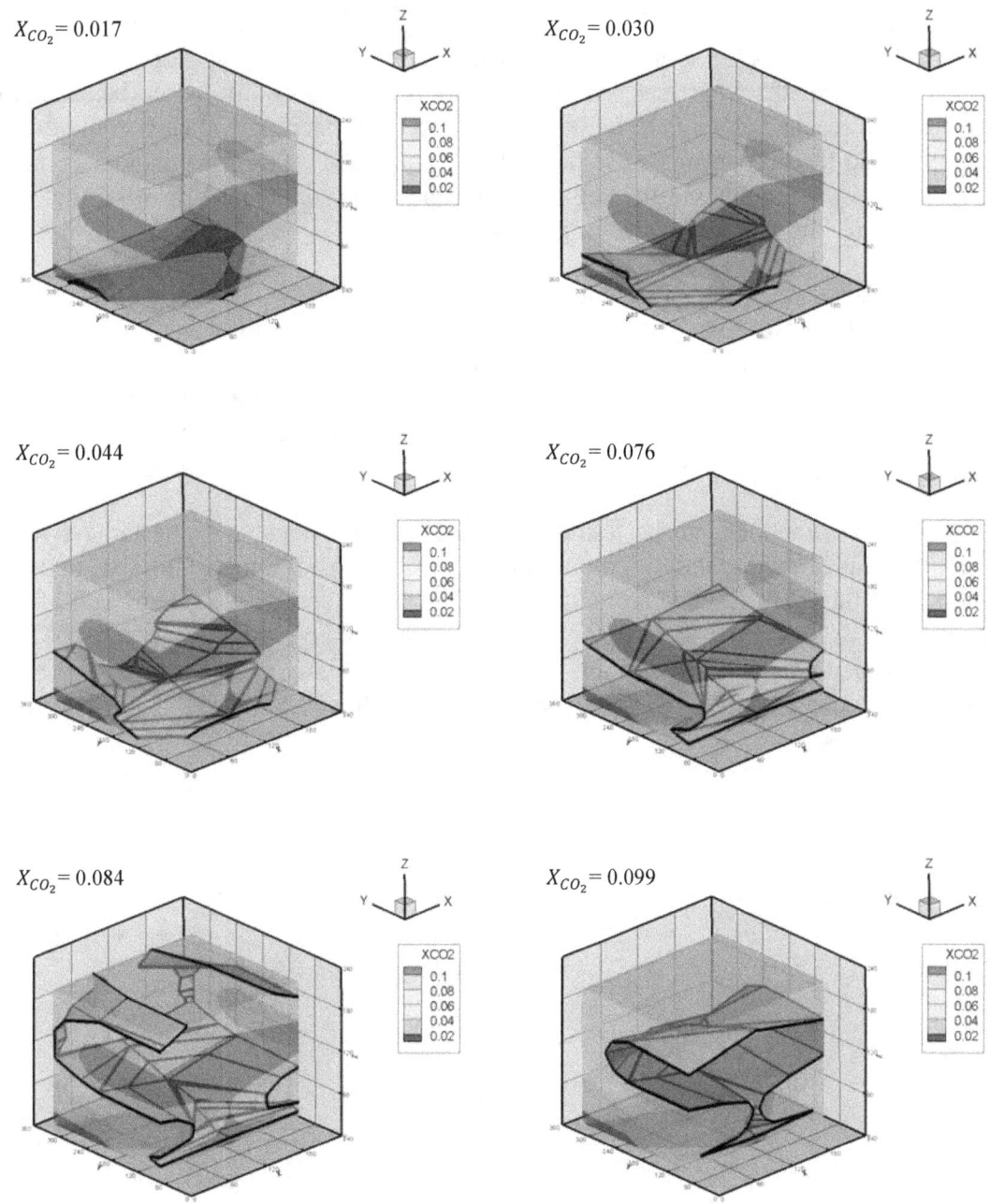

Figure 3.21: Multiple views of the interior carbon dioxide volume fraction contour plot for the ethanol experiments depicting isosurfaces (indicated by black lines) at various specific volume fractions

53

Considering that the O_2 volume fraction was generally low throughout the room, the CO volume fraction was a second indication of the extent that the room was underventilated. Insufficient O_2 causes incomplete formation of CO_2 resulting in CO. The CO volume fractions are displayed in Figure 3.22 for heptane and Figure 3.23 for ethanol. For both fuels, the largest volume fractions of CO occurred near the floor just behind the burners. The only difference being that heptane volume fraction reached about 0.16 and ethanol about 0.18. The highest carbon monoxide levels observed beyond the burner area were 0.049 for heptane and 0.064 for ethanol. In the heptane plot, this value was found along the ceiling centerline. For the ethanol fires, on the other hand, these values were found in the rear ceiling corners.

Figure 3.24 presents the H_2O vapor volume fraction for heptane estimated from the CO and CO_2 species volume fractions. The highest levels of H_2O volume fractions were near the rear of the burner reaching 0.18. High levels were also found from 60 cm above the floor to the ceiling along the front wall, reaching concentrations of 0.14. Figure 3.25 shows the H_2O volume fraction for ethanol. The upper half of the compartment had large volume fractions of about 0.20, but the highest values were again on the rear side of the burner reaching about 0.30.

The THC volume fractions can be found in Figure 3.26 for heptane. The largest values were seen near the rear of the burner with maximum concentrations of about 0.07. Beyond the burner area the rest of the room reached a maximum of 0.02 in the center of the rear wall. Figure 3.27 displays the THC volume fractions for ethanol. The volume fractions reached a maximum around 0.05 near the rear of the burner. The largest level reached in the rest of the room was about 0.02 located at the edges where the ceiling meets both side walls.

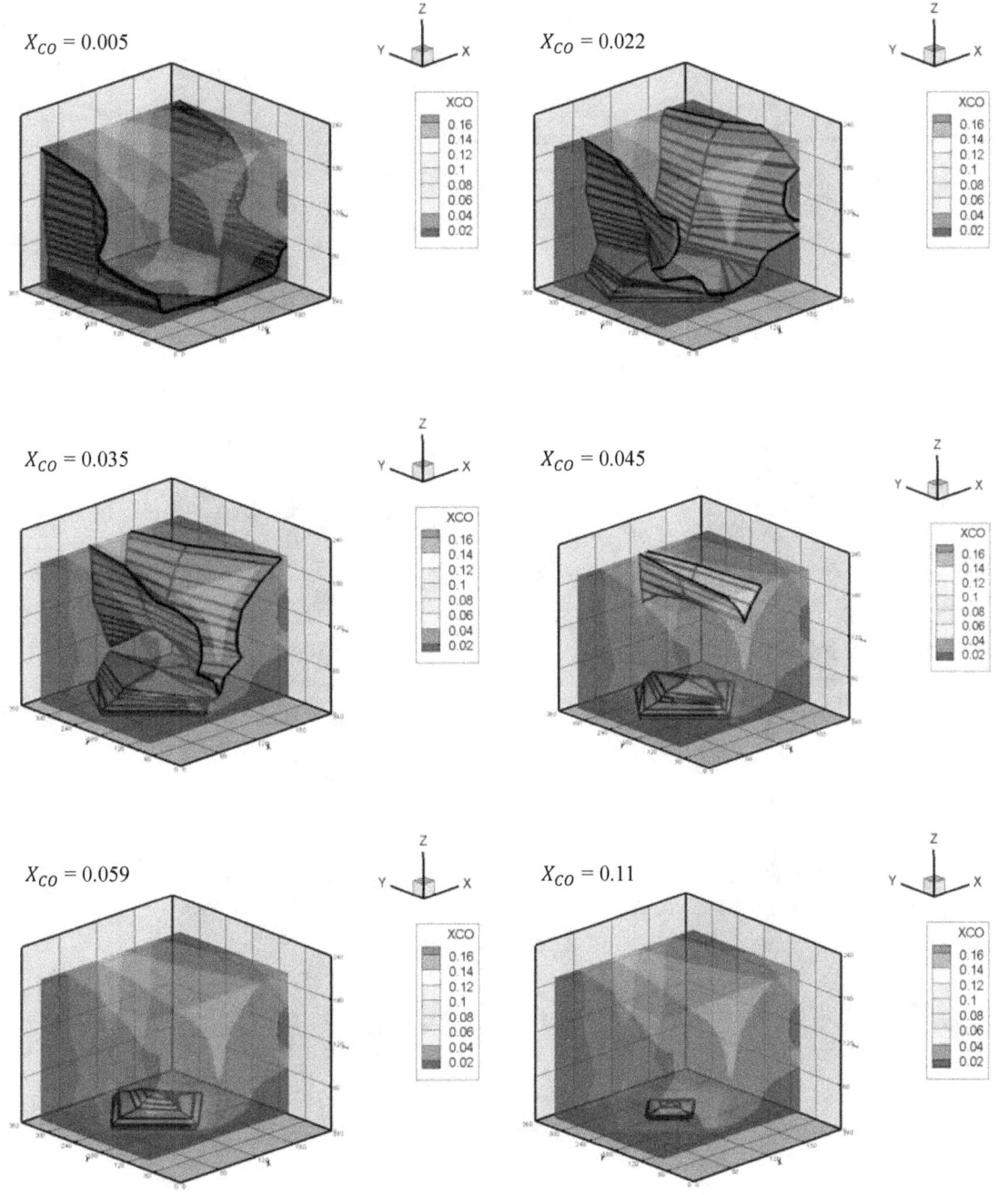

Figure 3.22: Multiple views of the interior carbon monoxide volume fraction contour plot for the heptane experiments depicting isosurfaces (indicated by black lines) at various specific volume fractions

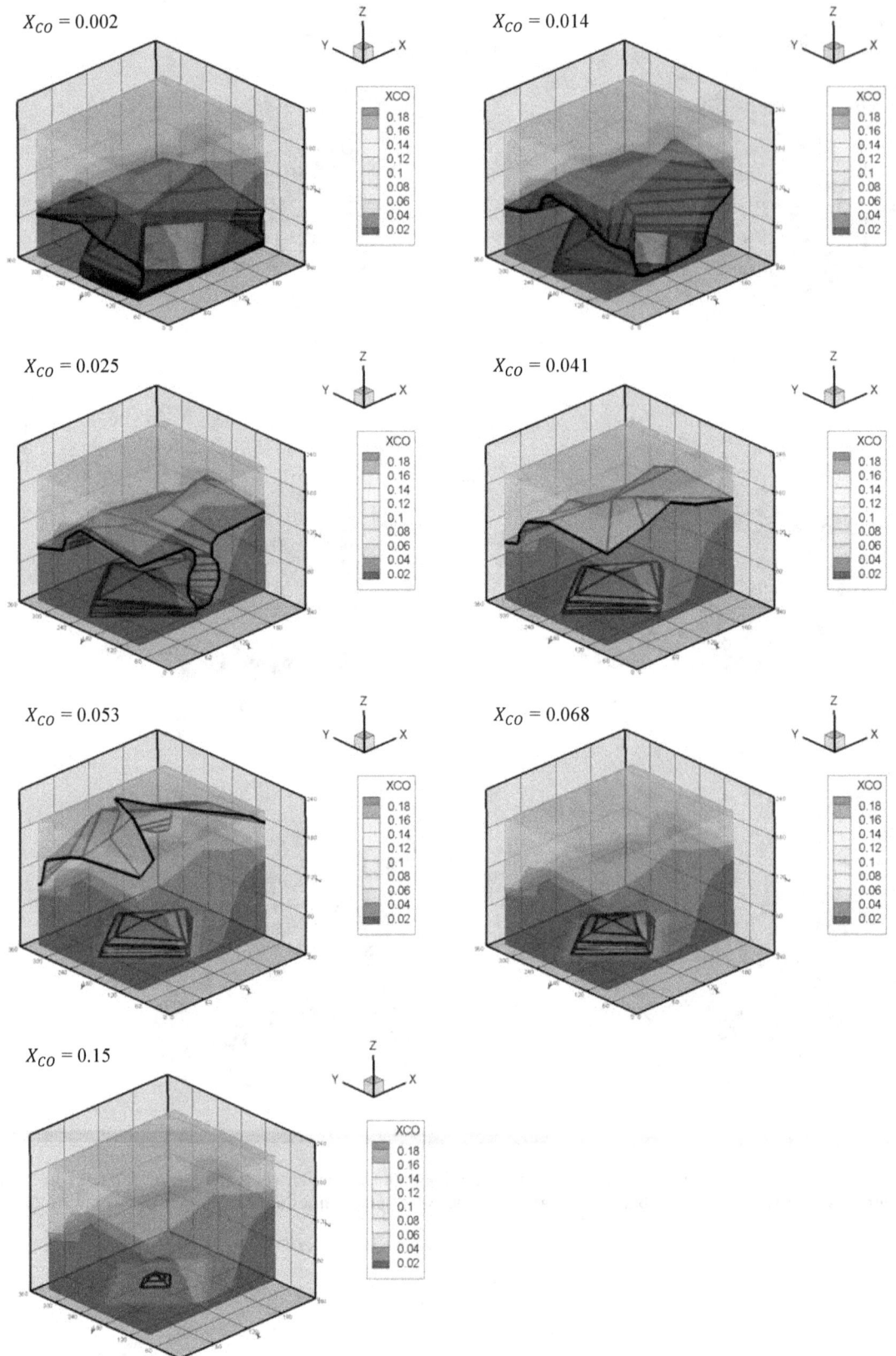

Figure 3.23: Multiple views of the interior carbon monoxide volume fraction contour plot for the ethanol experiments depicting isosurfaces (indicated by black lines) at various specific volume fractions

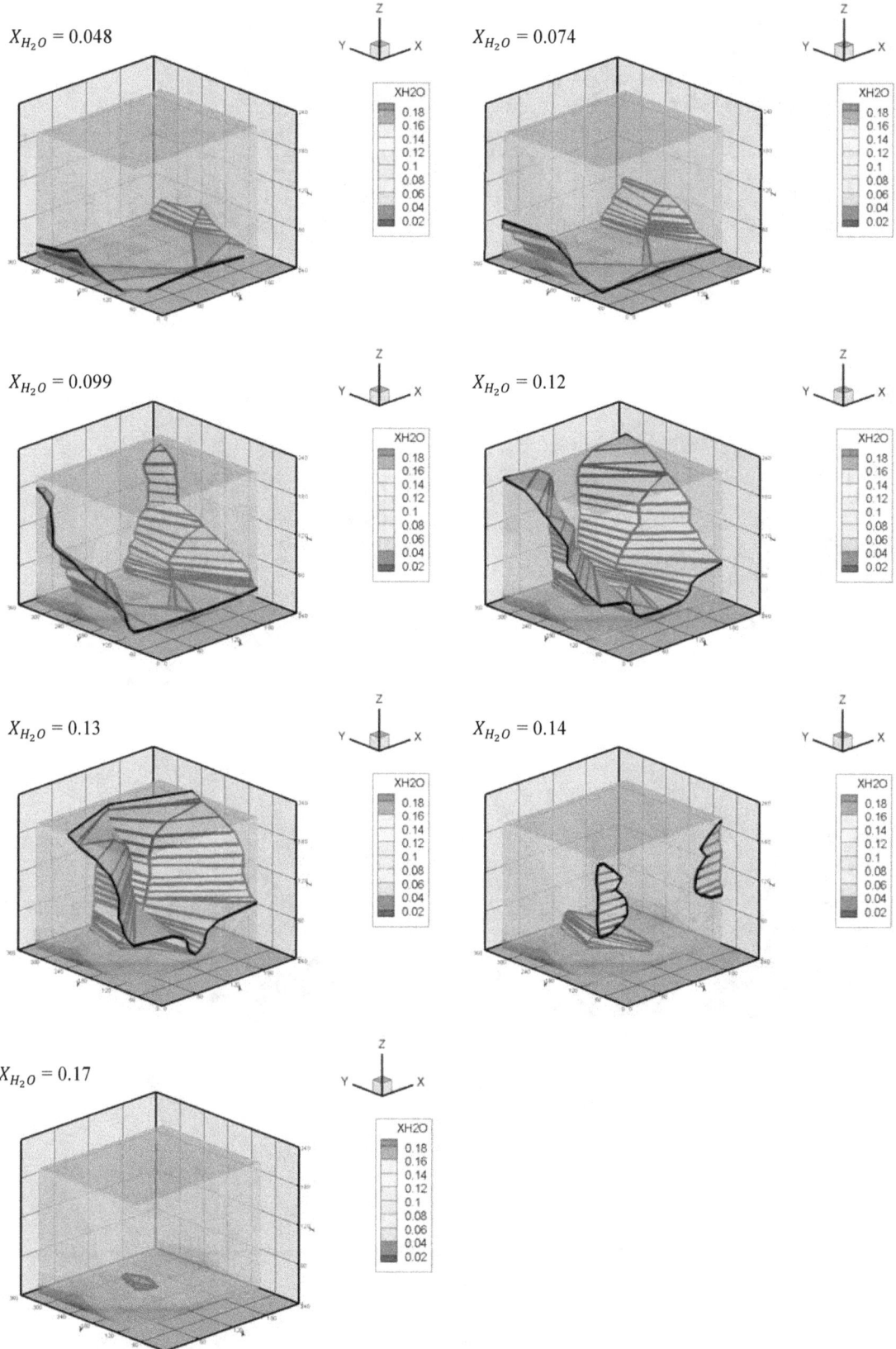

Figure 3.24: Multiple views of the interior gaseous water volume fraction contour plot for the heptane experiments depicting isosurfaces (indicated by black lines) at various specific volume fractions

57

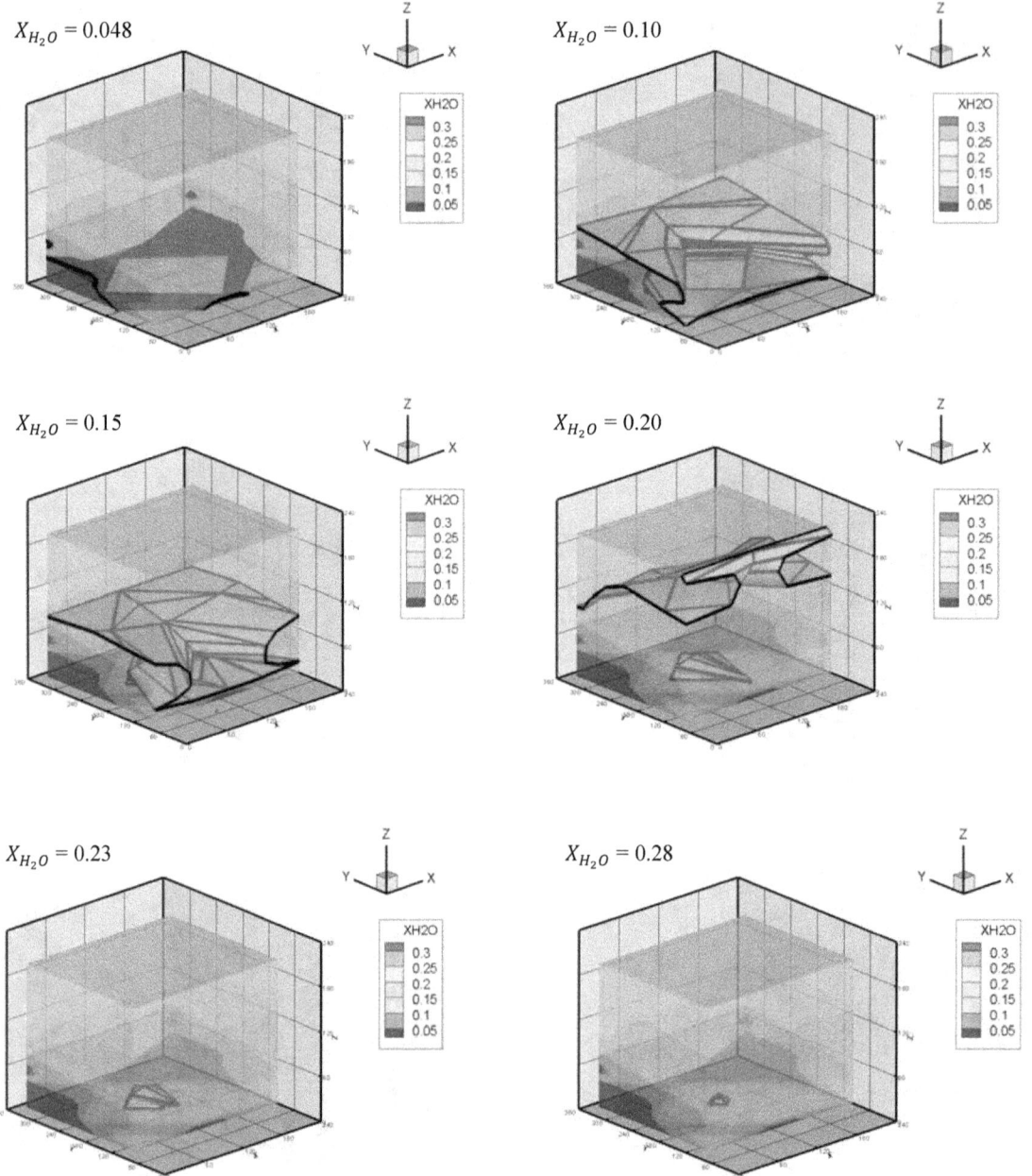

Figure 3.25: Multiple views of the interior gaseous water volume fraction contour plot for the ethanol experiments depicting isosurfaces (indicated by black lines) at various specific volume fractions

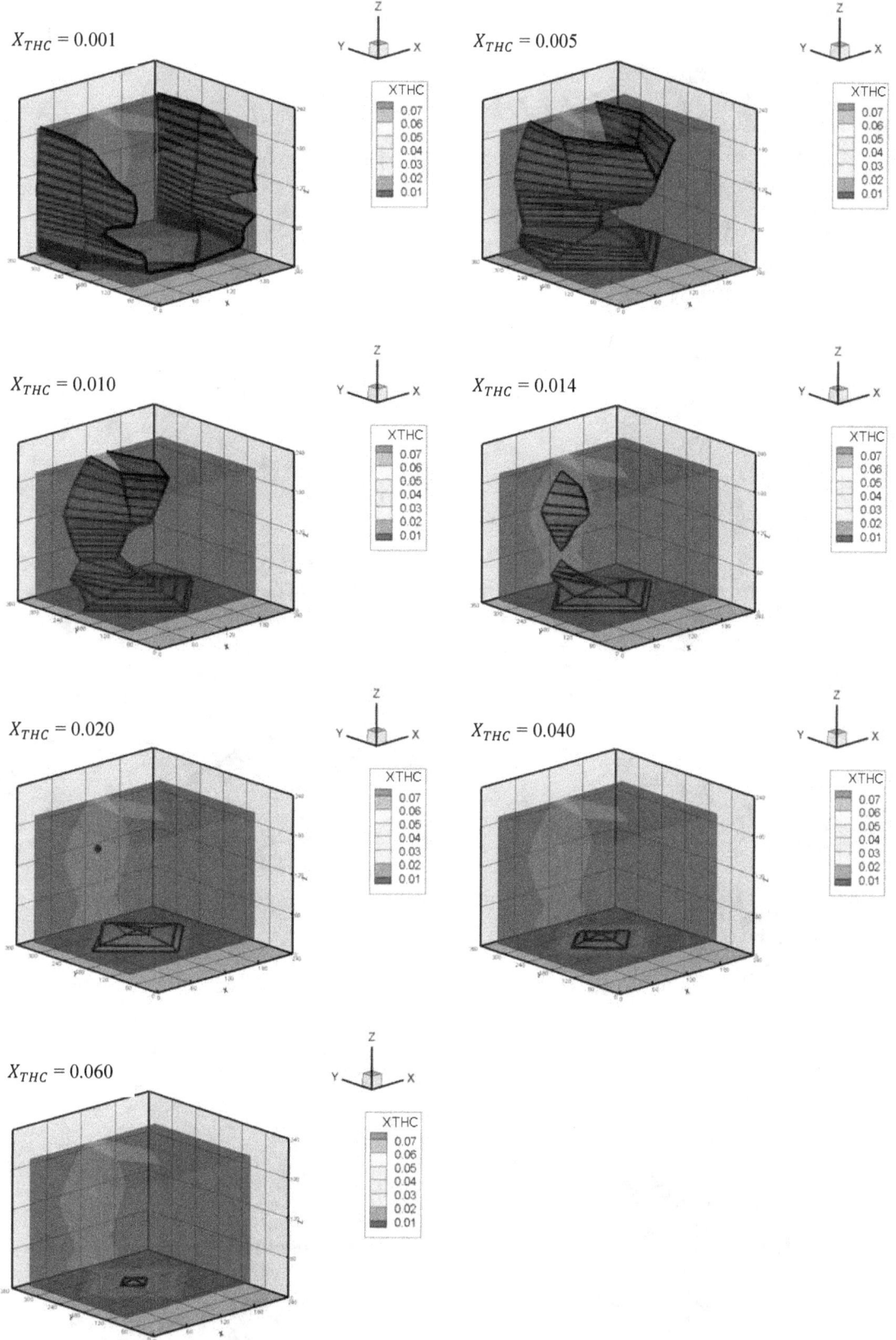

Figure 3.26: Multiple views of the interior total hydrocarbon volume fraction contour plot for the heptane experiments depicting isosurfaces (indicated by black lines) at various specific volume fractions

59

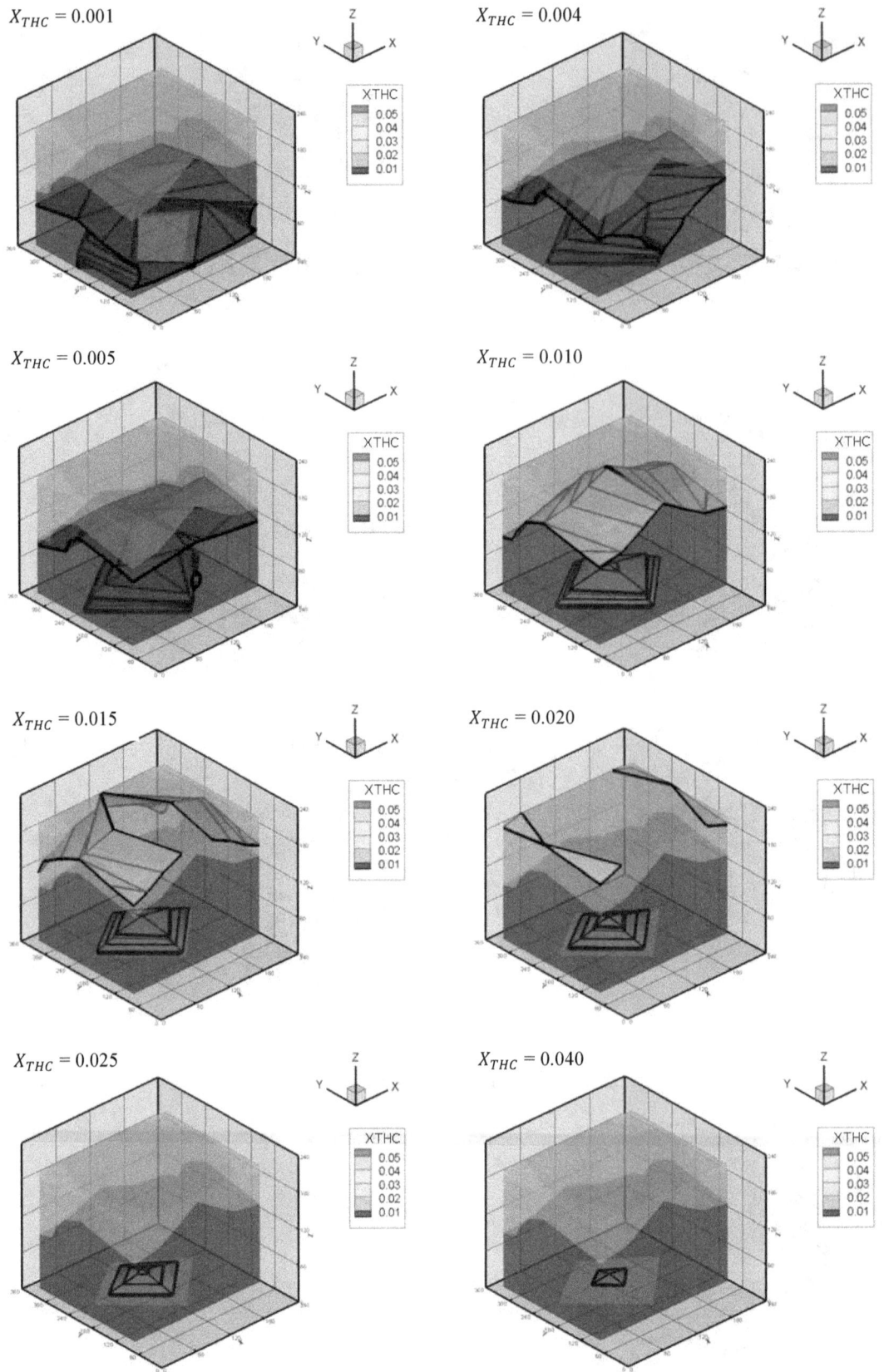

Figure 3.27: Multiple views of the interior total hydrocarbon volume fraction contour plot for the ethanol experiments depicting isosurfaces (indicated by black lines) at various specific volume fractions

60

3.6 Gravimetric Soot

Soot samples were collected during 1 min to 5 min sample times after the heat release rate was quasi-steady for many of the experiments. Measurements were conducted by gravimetric soot probes described in Sec. 2.2.3. Figure 3.28 and Figure 3.29 present the steady state gravimetric soot mass fraction measurements for the front and rear sample locations, respectively, as a function of the measured heat release rate. The maximum soot mass fractions reached 0.04 for the heptane and 0.007 for the ethanol fires. The species mass fraction results are examined further in Sec. 4.1 of this report.

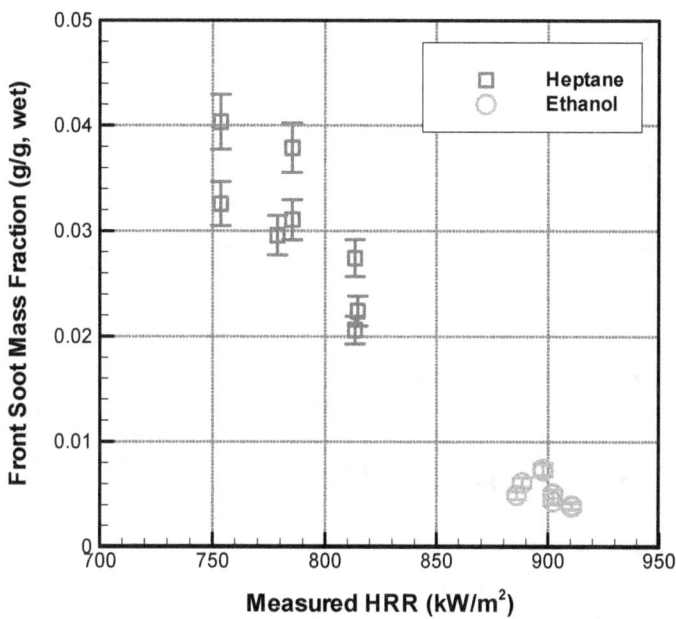

Figure 3.28: Steady state gravimetric soot mass fraction measurements at the front sample probe location

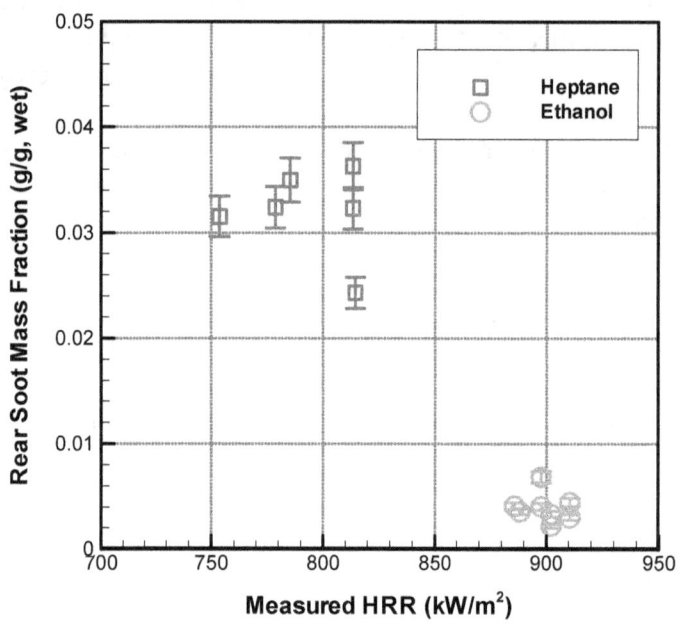

Figure 3.29: Steady state gravimetric soot mass fraction measurements at the rear sample probe location

4 COMPARTMENT CHEMISTRY ANALYSIS

4.1 Mixture Fraction Analysis

It is important to consider the compartment fire composition measurements in terms of the mixture fraction. The use of mixture fraction to analyze flame data was first used by Bilger [37] and later modified by Peters [38] and others. The mixture fraction approach has been widely used to represent the chemistry in turbulent flame models and fire field models, and has been used to analyze the structure of laminar counterflowing and coflowing hydrocarbon and alcohol flames [39,40].

Pool fires and compartment fires differ from simple laminar flames, as they are typically transient and turbulent by nature. Yet, application of the mixture fraction concept to these complex combustion situations can provide additional insight into the fire. The mixture fraction approach allows evaluation of a set of species measurements in terms of self-consistency, and at the same time facilitates rapid assessment of the overall behavior of a combustion system. Floyd et al [41] applied the mixture fraction approach to evaluate the species composition at various locations in compartment fires. Pitts [42] measured the local equivalence ratio at various locations in compartment fires, investigating the possibility of a correlation for CO. Since there is a one-to-one correspondence between mixture fraction and equivalence ratio, the approach used here is similar to that used previously by Pitts [42] and other experimentalists, with the difference that soot is considered in the analysis of mixture fraction and local equivalence ratio.

Sivathanu and Faeth [43] considered the relationship between soot and mixture fraction in an effort to improve the understanding associated with radiative emissions from fires. Their measurements [43] clearly showed that soot did not correlate well with mixture fraction in laminar hydrocarbon diffusion flames. Their data suggest, however, a relationship between soot volume fraction and temperature in the fuel rich regions of turbulent hydrocarbon diffusion flames.

Recently, a mixture fraction analysis was performed to investigate the characteristics of chemical species production in the upper layer of the 2/5 scale compartment based on the ISO-9705 room [5]. The analysis showed that plotting the local composition as a function of the mixture fraction collapsed hundreds of species measurements from an assortment of compartment conditions, with varying heat release rates, burner types and spatial locations, into a few coherent lines or bands. Also, inclusion of soot into mixture fraction analysis allowed identification of fuel rich or under-ventilated conditions for the compartment fires of smoky fuels, such as heptane, toluene, and polystyrene. The analysis performed here for the full-scale experimental data is an extension of our previous study [5] for the reduced-scale compartment fires.

In this section, the significance of the inclusion of soot as part of the mixture fraction analysis was investigated. The importance of measurement uncertainty is highlighted, and its value is quantified as part of the mixture fraction analysis. An explanation of how the mixture fraction and uncertainty were determined can be found the previous full scale report [6].

4.2   Species Composition Results in terms of Mixture Fraction

The gas species mass fractions were measured by gas analyzers on a dry basis. The gas species mass fractions were converted to wet mass fractions. The mass fraction of H_2O was not measured in the experiments; the value of this species in this report (shown in Figure 4.1 through Figure 4.4) is estimated from the stoichiometric relation found Ref [6]. The mass fractions of the unburned hydrocarbons (UH) in each plot were taken from the hydrocarbon analyzer measurements.

The gravimetric soot sample and the gas species mass fractions were not measured at the same position. The gravimetric sampling system took measurements at the front ($x = 189$ cm, $y = 25$ cm, $z = 208$ cm) and rear ($x = 189$ cm, $y = 286$ cm, $z = 208$ cm) of the compartment. Measurements of the gas species mass fractions were conducted at various heights at $y = 85$ cm, 225 cm, and 330 cm for $x = 10$ cm and 120 cm. In order to obtain the species data at the points at which the soot was sampled, linear interpolation was performed on data measured to determine the values at $z = 200$ mm. The species data are considered in terms of the species mass fraction (Y_i), which is plotted as a function of the local mixture fraction (Z), based on the fuel mass.

The lines in Figure 4.1 through Figure 4.4 represent complete stoichiometric combustion and represents the ideal case when only CO_2 is produced (no CO or soot). At any single location, the mixture fraction can vary from lean to rich, due to the dynamics of the fire. The stoichiometric mixture fraction (Z_{st}) is a useful reference point for consideration of fire chemistry. For fuel lean conditions ($Z < Z_{st}$), the measured mass fractions of unburned fuel are near zero. As the mixture fraction increases, the mass fraction of oxygen decreases, and the carbon dioxide and water vapor mass fractions increase. For heptane and ethanol mixture fraction values greater than stoichiometric, the oxygen mass fraction approaches zero, whereas the fraction of unburned fuel increases approximately linearly. Soot is shown only in Figure 4.2 and Figure 4.4 and is presented with the time-averaged gas species as a function of mixture fraction considering soot.

In some of the experiments, soot was collected twice resulting in more data points in Figure 4.2 and Figure 4.4 than Figure 4.1 and Figure 4.3.

Figure 4.1 and Figure 4.2 present the mass fractions of gas species measured during the ethanol experiments in both the front and rear position of the compartment as a function of mixture fraction. The figures show the time-averaged steady-state measurements without and with soot, respectively. Since the experiment was operated in an under-ventilated condition, all of the species data were in the fuel rich regime ($Z > Z_{st} = 0.1$). The following conclusions are made based on the narrow range of data collected and may not necessarily hold true over a larger range. In Figure 4.1 the ideal burning line over predicts the CO_2 mass fraction in the ethanol fire by about 28%, but the ideal burning line under predicts the H_2O mass fraction by about 15%. Due to the depletion of oxygen, significant CO concentrations are present, and the concentrations increase with a similar slope to the ideal burning line as the mixture fraction increases. Even for fuel rich condition, ethanol fires do not produce much soot or unburned hydrocarbons, leaving CO as the main cause for variations from the ideal burning lines. Therefore, the simple traditional mixture fraction approach for CO_2 correlates well to the experimental results for the sum of both CO_2 and CO. In Figure 4.2, the mixture fractions calculated with soot are almost the same as those without soot because amounts of soot are very small. However, the gas species in the averaged mixture fraction of $Z = 0.13$ in Figure 4.1 move to $Z = 0.14$ in Figure 4.2. The averaged soot mass fraction of these points is about .0044 g/g.

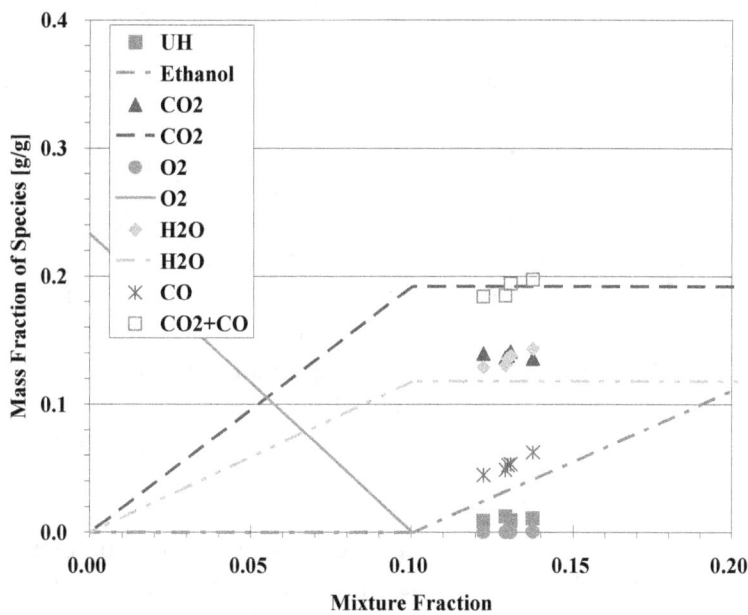

Figure 4.1: Mass fractions of front and rear compartment gas species for the ethanol experiments ISOEth42, ISOEth47, and ISOEth48 time-averaged measurements as a function of mixture fraction without soot

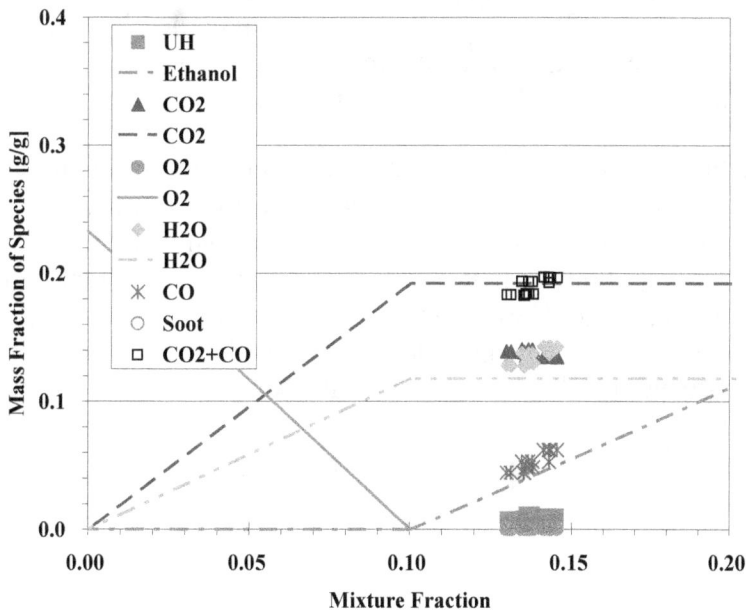

Figure 4.2: Mass fractions of front and rear compartment gas species for the ethanol experiments ISOEth42, ISOEth47, and ISOEth48 time-averaged measurements as a function of mixture fraction including soot

Figure 4.3 and Figure 4.4 shows the mass fraction as a function of mixture fraction for the fires burning heptane. The species mass fractions were measured when the HRR was quasi-steady and in under-ventilated conditions, indicated by near zero O_2 values and exterior burning, creating an interior atmosphere that is fuel-rich.

Figure 4.3 is reported without consideration of soot in the definition of mixture fraction. In some cases, like ethanol or any well-ventilated fire, where the inclusion of soot has relatively little influence over the mixture fraction, soot can be disregarded. But in other cases like heptane where more soot is produced, the effects of soot on the mixture fraction are more profound. The heptane plot quantitatively changes when soot is considered. Figure 4.4 shows that inclusion of soot shifts and stretches the range of mixture fraction. A large amount of the unburned fuel was shown to be converted to be soot. After soot is included, the slope for the sum of the soot and unburned hydrocarbons (UH) appears to closely follow the idealized mixture fraction line for the unburned fuel.

In our previous work, it was shown that the measured chemical species matched the idealized mixture fractions well for fuel-lean conditions and deviated from the idealized mixture fractions in fuel-rich conditions, particularly more so for heavier sooting fuels [6]. The results shown in Figure 4.1 through Figure 4.4 verify and extend the results of the previous experiments illustrating the failings of the ideal mixture fraction relationship for predicting the real behavior of gas species in fuel-rich environments. In the earlier work a large amount of the unburned hydrocarbon was converted to soot [6][1]. Also the sum of the unburned hydrocarbons and soot volume fractions was shown to correlate well with the ideal volume fraction of the unburned hydrocarbons [6]. That same behavior can be observed in Figure 4.4.

Figure 4.5 shows the relationship between the mixture fractions with and without soot for the ethanol and heptane fires. As expected, the mixture fractions with soot in the ethanol fires show little difference from the ones without soot. The uncertainty is smaller for ethanol because the ethanol fueled fires show less variation in HRR. When the compartment is fuel rich, the mixture fractions in the heptane fires vary significantly depending on whether the calculations are made with or without soot. Similar variance magnitudes were found in the fuel-rich zone of the heptane fires in the previous experiments [6] as are evident in Figure 4.5.

---

[1] In the figures of Section 4.1.3 and 4.1.4 of NIST Technical Note 1603 the lines for UH+soot are erroneously labeled as CO+soot.

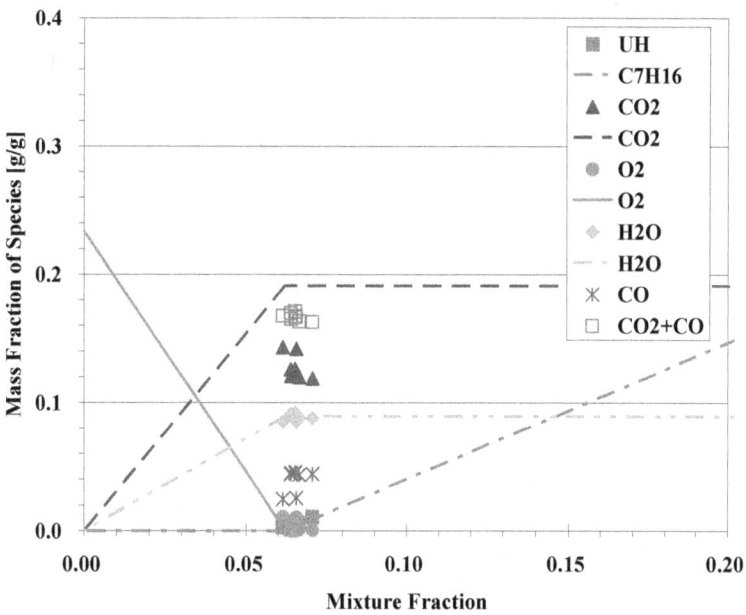

Figure 4.3: Mass fractions of front and rear compartment gas species for the heptane experiments ISOHept38, ISOHept39, ISOHept46, and ISOHept51 time-averaged measurements as a function of mixture fraction without soot

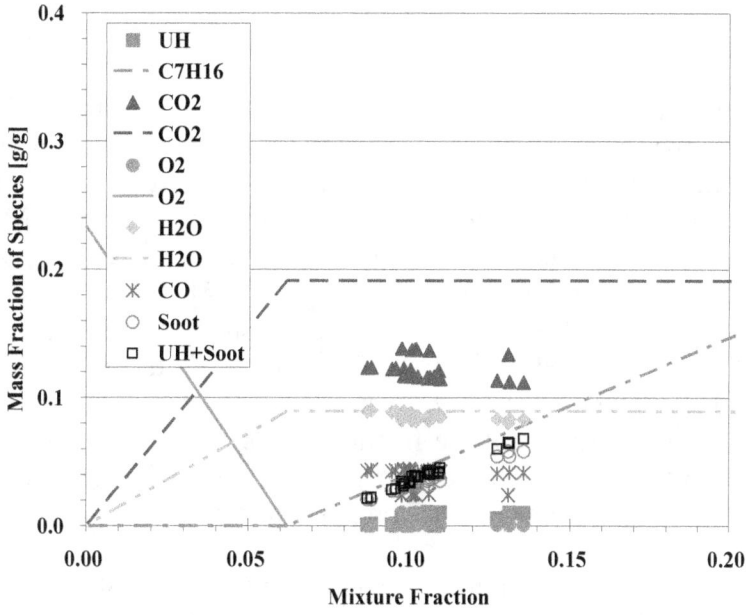

Figure 4.4: Mass fractions of front and rear compartment gas species for the heptane experiments ISOHept38, ISOHept39, ISOHept46, and ISOHept51 time-averaged measurements as a function of mixture fraction including soot

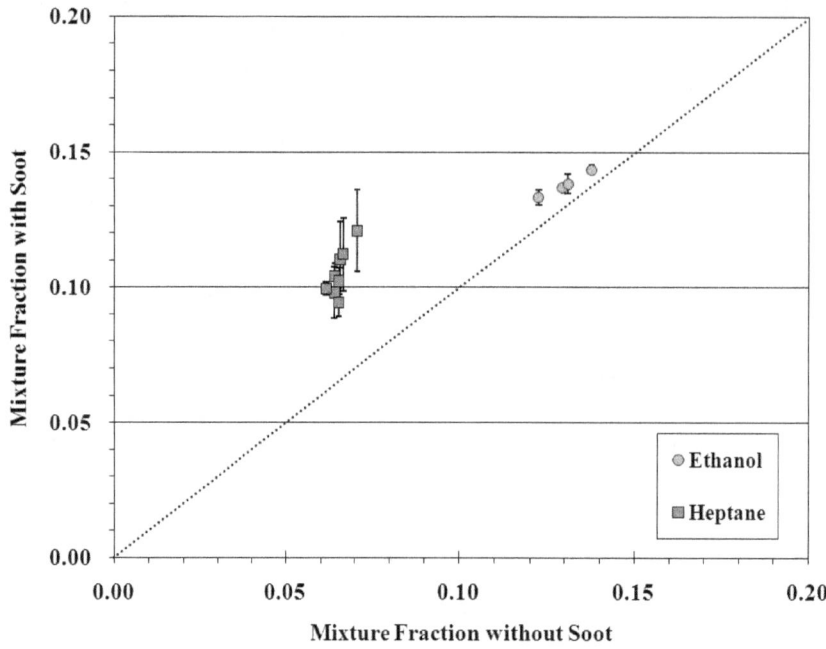

Figure 4.5: Comparison of mixture fraction calculated with and without soot using the time-averaged species measurements when the HRR was quasi-steady

4.3 Carbon Balance

Compartment measurements show that elemental carbon was primarily distributed among soot, CO_2, CO, and THC in the upper layer of the compartment. Soot was not measured at all of the 3D points in these experiments because of the labor intensive soot extraction method. Instead, soot only was measured at two points in the front and rear of the compartment near the ceiling. The corresponding gas samples were taken by the moving probes at the closest location to the soot measurement probes. In our previous reduced-scale study [5], the fractional mass-based amount of carbon was used to analyze the species compositions. This parameter has an advantage that the values are bounded from 0 to 1, contrary to the production yields or generation factor (defined below) which have been typically used to present composition results. Our previous study [5] showed the trends of the fractional mass-based amount of carbon were very similar in appearance to those of the production yields or generation factor, indicating the new parameter is a reasonable way to represent the composition results.

This fractional mass-based amount of carbon that exists in the form of carbon monoxide (F_{CO}) or carbonaceous soot (F_{soot}) is related to the mass fractions of carbon containing species at each measurement location as:

$$F_{soot} = \frac{Y_{soot}}{\frac{12}{16}Y_{CH_4} + \frac{12}{44}Y_{CO_2} + \frac{12}{28}Y_{CO} + Y_{soot}} \; ; \; F_{CO} = \frac{\frac{12}{28}Y_{CO}}{\frac{12}{16}Y_{CH_4} + \frac{12}{44}Y_{CO_2} + \frac{12}{28}Y_{CO} + Y_{soot}} \qquad 1$$

68

In the results presented for the compartment data, the value of X_s, which is a representation of the amount of carbonaceous soot is defined as:

$$X_s = \frac{Y_{soot}}{MW_C} \bigg/ \sum \frac{Y_i}{MW_i}$$

2

Table 4.1 lists F_{soot} and F_{CO} based on averages of the quasi-steady species measurements at the front and rear locations for the heptane and the ethanol fires. For convenience, the heat release rate (HRR), the local equivalence ratio (ϕ), and the F_{CO}/F_{soot} ratio are also included in the table. Figure 4.6 and Figure 4.7 show F_{CO} and F_{soot} as functions of the local equivalence ratio for the data from this set of experiments (in color) and the data from the previous report (in black) [6]. The F_{CO} maximum in the front was 0.24 for heptane and 0.37 for ethanol. In Figure 4.6 the difference between the rear and front values for heptane can be clearly seen. With the exception of the last two HRRs, all of the rear F_{CO} values are about 0.03. This indicates that less carbon is being produced in the form of CO in the rear of the room than in the front of the room. However, ethanol has less variance in its values and does not distinctly show a difference in values between the rear and front of the room. Overall, the new heptane data fell in line with the previous heptane data scatter. The values from ethanol were at the upper bounds of the distribution.

The F_{soot} was largest for the heptane fires, reaching a value of 0.57. This means that in those cases, about 57 % of the carbon existed in the form of soot. The F_{soot} in the ethanol experiments reached a maximum of 0.10. When compared to the previous data, the heptane results fell in line with the previous results. However, the ethanol values were all located at the fringe of what could be considered the lower boundary of the scatter.

In the richest heptane fire (ϕ=2.2), the sum of F_{CO} and F_{soot} reached 0.70, indicating that about 70 % of the carbon existed in the form of CO or soot, with relatively little carbon in the form of CO_2 or unburned fuel. The maximum values of the sum of F_{CO} and F_{soot} were 0.70 and 0.41 for heptane and ethanol, respectively. Table 4.1 lists value of F_{CO}/F_{soot}, which depends on fuel type, and physical location. Its value was less than 1.0, indicating more soot than CO, for all of the heptane experiments and greater than 1.0, indicating more CO than soot, for the ethanol experiments.

In our previous report on measurements in the ISO room it was found that the values of F_{soot} were different for the different fuels and tended to increase with the local equivalence ratio (or mixture faction) [6]. A similar increase can be seen in the heptane results in Figure 4.7. However, there is too little variance in the ethanol equivalence ratio values to see a pattern. A difference worth noting between the results here and the previous report is that ethanol has a high F_{CO} and a particularly low F_{soot}, resulting in a much higher F_{CO}/F_{soot} value than any other fuel used in this series of experiments or the last series. None of the fuels tested in the previous report showed that behavior. The difference between ethanol and all of the other fuels is that it was the only oxygenated fuel used during this series of reports [5, 6].

Table 4.1: Average fractional soot, CO and CO/soot ratio at the front and rear compartment measurement locations

Fuel	HRR [kW]	Rear				Front			
		ϕ_{local}	F_{CO}	F_{soot}	F_{CO}/F_{soot}	ϕ_{local}	F_{CO}	F_{soot}	F_{CO}/F_{soot}
Heptane	754	2.088	0.023	0.571	0.040	1.861	0.151	0.438	0.345
	754	1.514	0.032	0.412	0.077	1.696	0.166	0.385	0.431
	779	1.532	0.031	0.419	0.074	1.633	0.172	0.361	0.476
	815	1.365	0.035	0.349	0.100	1.484	0.189	0.299	0.634
	813	1.614	0.030	0.448	0.066	1.447	0.194	0.281	0.690
	813	1.530	0.031	0.419	0.075	1.587	0.177	0.344	0.515
	785	1.802	0.237	0.391	0.606	1.824	0.171	0.419	0.407
	785	2.224	0.193	0.505	0.382	1.679	0.185	0.370	0.500
Ethanol	886	1.419	0.295	0.058	5.128	1.395	0.301	0.070	4.306
	888	1.408	0.298	0.051	5.894	1.421	0.296	0.086	3.452
	902	1.379	0.304	0.032	9.542	1.383	0.303	0.062	4.887
	902	1.403	0.299	0.048	6.281	1.399	0.300	0.072	4.167
	911	1.591	0.304	0.057	5.343	1.492	0.303	0.053	5.742
	911	1.559	0.370	0.039	9.403	1.489	0.328	0.051	6.408
	898	1.479	0.284	0.093	3.064	1.447	0.291	0.100	2.893
	898	1.418	0.296	0.057	5.205	1.449	0.290	0.102	2.851

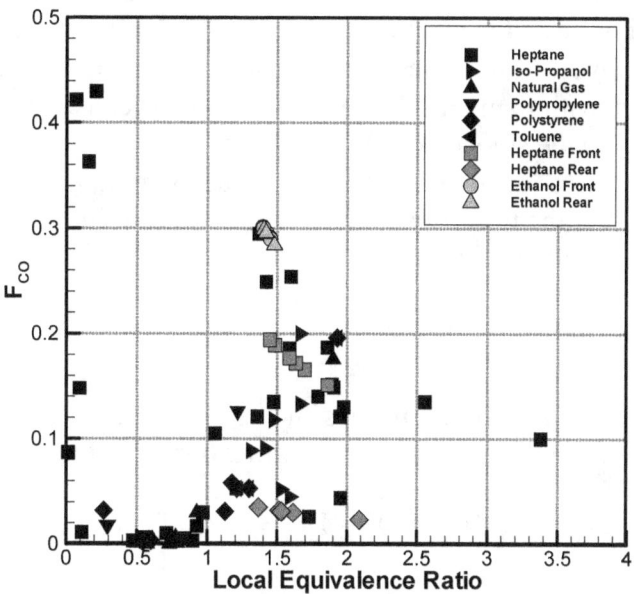

Figure 4.6: The values of F_{co} as a function of the local equivalence ratio for the time averaged measurements during the period when the HRR was quasi-steady. Data points in color are from this study. Data points in black are from the analogous full-scale full door width experiments from Ref. [6].

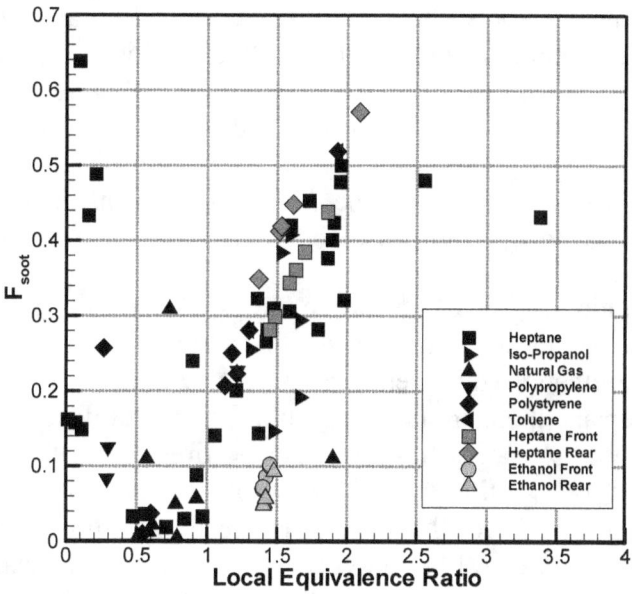

Figure 4.7: The values of F_{soot} as a function of the local equivalence ratio for the time averaged measurements during the period when the HRR was quasi-steady. Data points in color are from this study. Data points in black are from the analogous full-scale full door width experiments from Ref. [6].

Measurements by Köylü et al [44] and Puri and Santoro [45] showed that there is a linear relationship between the emission of soot and CO from buoyant turbulent diffusion flames burning various hydrocarbon fuels (acetylene, propene, etc.). Measurements in the fuel lean (overfire) plume region of hydrocarbon fires showed that the soot and CO generation factors (η_S and η_{CO}) tended to increase with flame residence time, until near-constant values were reached after long times (compared to the smoke point). Köylü et al [44] reported that the ratio of the CO and soot generation factors for a range of fuel types was such that, $\eta_{CO}/\eta_S = 0.34 \pm 0.09$. The generation rate was defined as the mass of soot (or gas species) produced per unit mass of fuel carbon consumed. This is slightly different than the soot (or gas species) yield (y_{CO} and y_S), which is based on the mass of all elements (not just carbon) in the fuel stream. The yield values were determined using the following equation.

$$y_i = Y_i / Z \qquad\qquad 3$$

The ratios of the yields and the generation rates, however, are equal, and their values can be determined at any location from the ratio of the mass fractions of CO and soot:

$$\eta_{CO}/\eta_{soot} = y_{CO}/y_{soot} = Y_{CO}/Y_{soot} = (7/3)\,F_{CO}/F_{soot} \qquad\qquad 4$$

The constant value (7/3) in Eq. 4 is the ratio of the total CO mass to the mass of carbon.

Table 4.2 lists y_{co}, y_{soot}, and the ratio y_{co}/y_{soot} based on the time-averaged species measurements at the front and rear compartment locations when the heat release rate was quasi-steady. The fire HRR and the local equivalence ratio are also listed. Much of the same data was used in Table 4.1. Figure 4.8 and Figure 4.9 show the yields of CO and soot as a function of the local equivalence ratio for heptane and ethanol, shown in color, and for the fuels used in the previous report, shown in solid black [6]. These figures are analogous to Figure 4.6 and Figure 4.7, with the parameters y_{CO} and y_{soot} considered in lieu of F_{CO} and F_{soot}. The trends and values of the data shown in the graphs were generally similar in appearance and consistent with the data presented in Table 4.1. The only difference was that the F_{CO} from the ethanol fires were at the upper boundary of the scatter while the y_{co} from the same fires fell into the middle of the scatter.

Figure 4.10 shows the ratio of the CO yield to the soot yield as a function of the local equivalence ratio for the same quasi-steady data shown in Figure 4.8 and Figure 4.9. While there was little variation in the ethanol equivalence ratio values, there were large variations in the y_{co}/y_{soot} ratio, while for heptane there are large variations in the equivalence ratio and little in the y_{co}/y_{soot} ratio. All of the y_{co}/y_{soot} ratios from the heptane fires fell in line with the past y_{co}/y_{soot} ratios while none of the y_{co}/y_{soot} ratios from the ethanol fires did. Figure 4.11 shows the CO yield as a function of the soot yield for the same quasi-steady data for the fires along with a line representing the results of Köylü et al [44]. Köylü reported about 30 % scatter in the ratio of the yields of CO to soot, which is considerably smaller than that seen in the figure. None of the points from this data set fell within the boundaries of this line. The values were mostly above this range, with the exception of the rear heptane values, which are below the range. When looking at the past data, none of the heptane points that fell below the Köylü et al range were yields taken in the rear of the compartment, but some points had fallen below the line [6]. The heptane y_{co}/y_{soot} ratios stated in the 2008 report were also considerably more scattered than the Köylü et al. reported range. The heptane results found here fall within the range reported in the previous report and all of the ethanol results were outside the scatter. Nevertheless, more data and analysis are needed to examine this relationship in the uppers layer of compartment fires, particularly for oxygenated fuels.

Table 4.2: Average yields of soot, CO and CO/soot ratio at the front and rear compartment measurement locations

Fuel	HRR [kW]	Rear				Front			
		ϕ_{local}	y_{CO}	y_{soot}	y_{CO}/y_{soot}	ϕ_{local}	y_{CO}	y_{soot}	y_{CO}/y_{soot}
Heptane	754	2.088	0.05	0.48	0.09	1.860	0.30	0.37	0.8
	754	1.514	0.06	0.35	0.18	1.695	0.33	0.32	1.0
	779	1.532	0.06	0.35	0.17	1.632	0.34	0.30	1.1
	815	1.365	0.07	0.29	0.23	1.483	0.37	0.25	1.5
	813	1.614	0.06	0.37	0.16	1.447	0.38	0.24	1.6
	813	1.530	0.06	0.35	0.17	1.530	0.35	0.29	1.2
	785	1.802	0.46	0.33	1.41	1.802	0.33	0.35	1.0
	785	2.224	0.38	0.42	0.89	2.224	0.36	0.31	1.2
Ethanol	886	1.419	0.36	0.03	11.96	1.395	0.37	0.04	10.0
	888	1.408	0.36	0.03	13.75	1.421	0.36	0.05	8.1
	902	1.379	0.37	0.02	22.26	1.383	0.37	0.03	11.4
	902	1.403	0.36	0.03	14.66	1.399	0.37	0.04	9.7
	911	1.591	0.44	0.03	14.91	1.492	0.40	0.03	14.5
	911	1.559	0.45	0.02	21.94	1.489	0.40	0.03	15.0
	898	1.479	0.35	0.05	7.15	1.447	0.35	0.05	6.8
	898	1.418	0.36	0.03	12.15	1.449	0.35	0.05	6.7

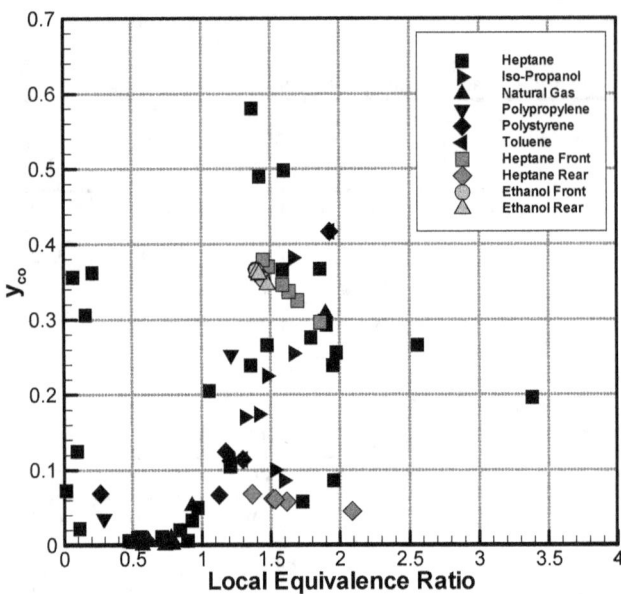

Figure 4.8: The CO yields as a function of the local equivalence ratio for the time averaged measurements during the period when the HRR was quasi-steady. Data points in color are from this study. Data points in black are from the analogous full-scale full door width experiments from Ref. [6].

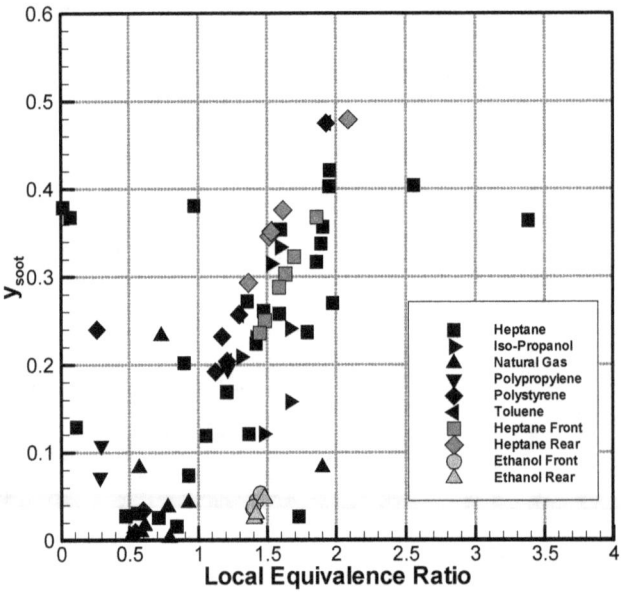

Figure 4.9: The soot yields as a function of the local equivalence ratio for the time averaged measurements during the period when the HRR was quasi-steady. Data points in color are from this study. Data points in black are from the analogous full-scale full door width experiments from Ref. [6].

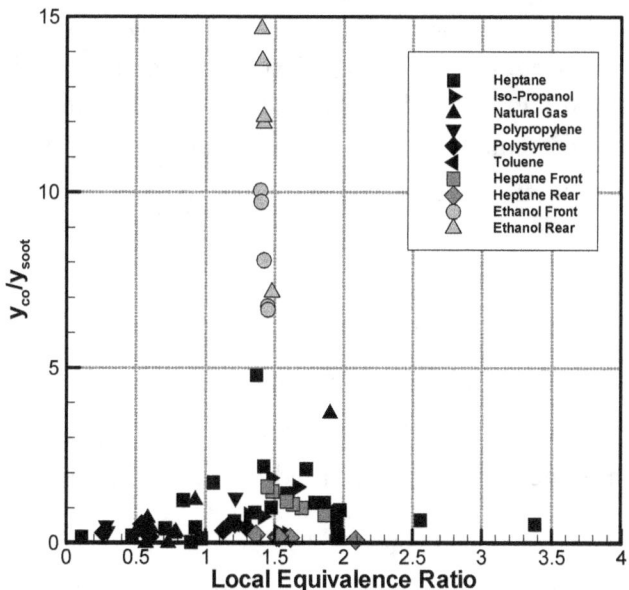

Figure 4.10: The ratio of the CO to soot yield as a function of the local equivalence ratio during the period when the HRR was quasi-steady. Data points in color are from this study. Data points in black are from the analogous full-scale full door width experiments from Ref. [6].

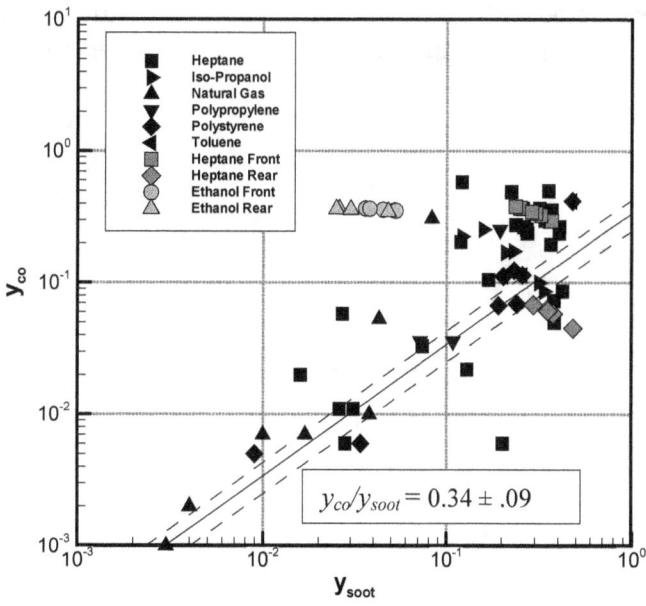

Figure 4.11: The CO yield as a function of the soot yield during the period when the HRR was quasi-steady [6]. Also shown is a line representing the results of Köylü [45] for the "over-fire" region of hydrocarbon pool fires. Data points in color are from this study. Data points in black are from the analogous full-scale full door width experiments from Ref. [6].

4.4 Combustion Efficiency

To better understand the compartment combustion chemistry, it is of interest to determine the combustion efficiency both in the exhaust stack and at various locations in the upper layer of the compartment. Moreover, the accurate prediction of burning fraction inside a compartment may provide useful information to understand the formations of CO and soot. The combustion efficiency (χ_a) is a global representation of the fractional amount of heat released by the fire as compared to complete combustion. It is defined as:

$$\chi_a = \frac{\Delta H_c}{\Delta H_{c,ideal}}$$

5

where $\Delta H_{c,ideal}$ is the net heat of complete combustion based on the conversion of all carbon and hydrogen in the fuel to CO_2 and H_2O (assumed to remain in the vapor phase) and ΔH_c is the net heat of combustion, which is the actual heat released in a chemical reaction. The value of χ_a is bounded by 0 % to 100 %.

The combustion efficiency was calculated in the exhaust stack based on the species measurements from the gas analyzers and was determined as the ratio of the HRR to the IHRR. A summary of averaged steady-state results of combustion efficiency in the exhaust stack is shown in Table 4.3. Figure 4.12 shows the combustion efficiency for both the heptane and ethanol fires based on measurements in the exhaust stack using measurements made during the steady-state burning periods as a function of the ideal heat release rate. As mentioned in Section 2.2, heat release rate measurements had a combined expanded relative uncertainty of 14 %. On average, the ethanol fires had a higher combustion efficiency than heptane.

Table 4.3: Summary of averaged steady-state results of combustion efficiency in the exhaust stack. The uncertainty, U, indicated here only reflects the statistical variation.

Experiment ID	IHRR (kW)		Combustion Efficiency (%)	
	Mean	*U*	*Mean*	*U*
ISOHept38	1008	1	74.8	3.6
ISOHept39	1000	1	77.9	3.5
ISOEth40	994	1	89.1	4.2
ISOEth42	1013	1	89.1	3.4
ISOHept45	994	5	81.9	4.3
ISOHept46	1011	3	80.4	3.3
ISOEth47	1000	1	91.1	4.1
ISOEth48	1001	1	89.7	3.2
ISOHept51	1001	1	78.5	3.5

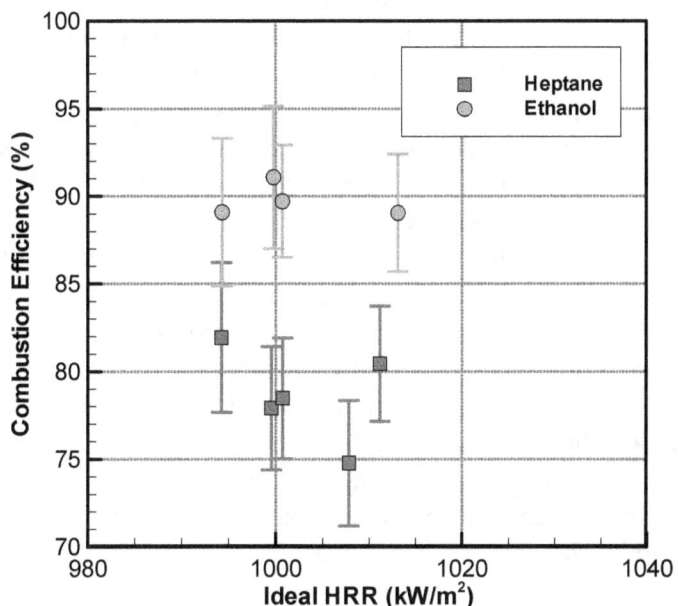

Figure 4.12: The combustion efficiency in the exhaust stack as a function of the ideal heat release rate

5 SUMMARY

This report documents a set of 9 full-scale ISO 9705 room under-ventilated fire experiments of detailed mapping of the interior dynamics of the room. Analysis was conducted on each case in order to verify the validity of results observed in the experiments.

Heptane and ethanol were burned in an ISO 9705 compartment with a 1/8 size door width (10 cm) in order to create underventilated fires. The fuels were sprayed into a 0.5 m^2 pan that was 0.1 m deep in order to maintain a steady heat release rate. This allowed for a long duration quasi-steady-state fire to be maintained. When the fires reached quasi-steady state, the measurements were as controlled as possible given the turbulent nature of the fire. Therefore, movable probes that measured temperature and gas species were able to gather data throughout the room in order to map the interior environment of the compartment. The heptane and ethanol experiments yielded maps of temperature and gas species.

The HRR and heat flux were measured in each experiment to show the reproducibility of the experiments for each fuel type and to establish the distinct differences between the two fuels. With the ideal HRR set to 1000 kW, the average measured HRRs varied from 754 kW to 815 kW for the heptane experiments and 886 kW to 911 kW for the ethanol experiments. The heat fluxes were generally higher at each location measured for ethanol than for heptane. Temperature measurements were taken at multiple locations throughout the compartment during the near steady state burning period to develop three dimensional contour plots for the different fuels. In reviewing the results of the temperature measurements for ethanol, it was found that the cooler temperatures in the compartment were in the front third of the compartment, near the door, and that the hottest temperatures existed in the rear two thirds of the compartment, particularly at the mid-height of the compartment. The heptane results revealed cool temperatures at the floor near the door and high temperatures in the rear upper corners of the compartment. The heptane experiments resulted in cooler temperatures extending further back into the compartment then the ethanol experiments, but hotter temperatures in the front upper half of the compartment. The heptane temperatures reached near 1300 °C while the ethanol only reached around 1100 °C. In addition to the movable temperature probes, two thermocouple arrays were hung in the front and the rear of the compartment. These arrays were used to verify the findings of the movable probes.

The gas species measurements were taken within the compartment by way of the same movable probes. For both fuels, the carbon dioxide volume fractions peaked in the areas between the oxygen peak and the carbon monoxide and total hydrocarbons peaks. For the case of ethanol, the gas species volume fraction plots showed relatively flat behavior between $z = 90$ cm and $z = 120$ cm. Heptane did not show the same behavior. The area around the burner experienced the highest extremes for both fuels. In this area, the oxygen and carbon dioxide volume fractions were low and the carbon monoxide, water vapor, and total hydrocarbons reached considerably higher peaks than in the rest of the compartment.

The species composition, mixture fraction, carbon balances, and combustion efficiency were calculated from the measured data. In the previous report, it was shown that for $Z < Z_{st}$ the mixture fraction model does a good job of predicting the chemical composition within the room [6]. For $Z > Z_{st}$ the agreement was not as good and in some cases really poor [6]. In these experiments, data was collected only for $Z > Z_{st}$. For the heptane experiments, the stoichiometric mixture fraction poorly predicted the chemical composition in the compartment while for the ethanol experiments it was found to predict the chemical composition much better, due to the minimal amounts of soot that ethanol produces.

For the heptane experiments, the carbon monoxide and soot yields and fractional amounts of carbon in the form of carbon monoxide and soot fell in line with the trends found in the previous report [6]. The soot yields and fractional amounts of carbon in the form of carbon monoxide and soot from the ethanol experiments were all on the outer limits of the scatter from the previous report. Only the carbon monoxide yields were definitely situated in the scatter recorded previously. It is also worth noting that when the yields were calculated and compared for ethanol, the y_{co}/y_{soot} ratio values versus the local equivalence ratios were well outside the of y_{co}/y_{soot} ratio values found in all of the fuels tested in the previous report due to high carbon monoxide yields and low soot yields. This behavior may be due to ethanol being the only oxygenated fuel used in this series of reports [5, 6]. With a door width that is the same as the one used here at the same HRR, the combustion efficiency for heptane was slightly higher in this report than the previous report. The ethanol experiment efficiency was about 90 % compared to the heptane experiment efficiency of roughly 80 %.

The data was collected in order to help improve the accuracy and reliability of computer fire models, in particular the FDS model developed by NIST. Among the various assumptions used in the development of previous versions of FDS, all chemical species were tied to the mixture fraction state relations. A single mixture fraction variable cannot be used for the prediction of carbon monoxide and soot, and the yield of these species was prescribed in FDS 4, rather than predicted. The yield of these species is usually not constant, but a complex function of their time-temperature history. This report provides computational fire modelers with information on the three dimensional temperature and gas species behavior as well as soot and mixture fraction data within an underventilated compartment for a sooty fuel and a minimally sooty fuel. The information can be used to guide the development and validation of fire models for underventilated compartment fires.

6 ACKNOWLEDGEMENTS

The authors wish to acknowledge the contributions of several people without whom this work would not be possible. The large fire lab staff, Lauren Delauder, Doris Reinhart, and Tony Chakalis did an excellent job of preparing and running the experiments as well as providing technical services to facilitate this test matrix. Kevin McGrattan and Jason Floyd provided invaluable insight and guidance into the design of the test matrix used here.

7 REFERENCES

1. *ISO9705 Fire Tests - Full-Scale Room Test for Surface Products First Edition*, International Organization for Standardization, Geneva, Switzerland, 1993

2. McGrattan, K. B., Baum, H., Rehm, R., Mell, W. and McDermott, R., *Fire Dynamics Simulator (Version 5): Technical Reference Guide*, 2010, **NIST SP 1018-5**.

3. Floyd, J. E. and McGrattan, K. B., *Validation of A CFD Fire Model Using Two Step Combustion Chemistry Using the NIST Reduced-Scale Ventilation-Limited Compartment Data*, 9th International IAFSS Symposium, Karlsruhe, Germany

4. National Fire Protection Association and Society of Fire Protection Engineering, *SFPE Handbook of Fire Protection Engineering*, National Fire Protection Association, Quincy, Mass, 2008.

5. Bundy, M., Hamins, A., Johnsson, E. L., Kim, S. C., Ko, G. H. and Lenhert, D. B., *Measurements of Heat and Combustion Products in Reduced-Scale Ventilation-Limited Compartment Fires*, 2007, **NIST TN 1483**.

6. Lock, A., Bundy, M., Johnson, E. L., Hamins, A., Ko, G. H., Hwang, C., Fuss, P. and Harris, R. H., *Experimental Study of the Effects of Fuel Type, Fuel Distribution, and Vent Size on Full-Scale Underventilated Compartment Fires in an ISO 9705 Room*, 2008, **NIST TN 1603**.

7. Beyler, C. L., *Major Species Production by Diffusion Flames in A 2-Layer Compartment Fire Environment*, Fire Safety Journal, 1986, **10 (1)** p. 47-56.

8. Zukoski, E. E., Morehart, J. H., Kubota, T. and Toner, S. J., *Species Production and Heat Release Rates in 2-Layered Natural-Gas Fires*, Combustion and Flame, 1991, **83 (3-4)** p. 325-332.

9. Brohez, S., *Uncertainty analysis of heat release rate measurement from oxygen consumption calorimetry*, Fire and Materials, 2005, **29 (6)** p. 383-394.

10. Bryner, N. P., Johnson, E. L. and Pitts, W. M., *Carbon Monoxide Production in Compartment Fires - Reduced-Scale Enclosure Test Facility*,1994, **NIST IR 5568**.

11. Pitts, W.M., E.L. Johnsson, and N.P. Bryner, *Carbon Monoxide Formation in Fires By High-Temperature Anaerobic Wood Pyrolysis.* Twenty-Fifth Symposium (International) on Combustion, 1994: p. 1445-1462.

12. Lattimer, B. Y. and Roby, R. J., *Carbon monoxide levels in structure fires: Effects of wood in the upper layer at a post-flashover compartment fire*, Fire Technology, 1998, **34 (4)** p. 325-355.

13. Lattimer, B. Y., Vandsburger, U. and Roby, R. J., *Species transport from post-flashover fires*, Fire Technology, 2005, **41 (4)** p. 235-254.

14. Gottuk, D. T., Roby, R. J. and Beyler, C. L., *The role of temperature on carbon monoxide production in compartment fires*, Fire Safety Journal, 1995, **24 (4)** p. 315-331.

15. Gann, R. G., Averill, J. D., Johnsson, E. L., Nyden, M. R. and Reacock, R. D., *Smoke component yields from room-scale fire tests*, 2003, **NIST TN 1453**.

16. Hirschler, M. M., *Analysis of work on smoke component yields from room-scale fire tests*, Fire and Materials, 2005, **29 (5)** p. 303-314.

17. Blomqvist, P. and Lonnermark, A., *Characterization of the combustion products in large-scale fire tests: Comparison of three experimental configurations*, Fire and Materials, 2001, **25 (2)** p. 71-81.

18. Pitts, W. M., *The Global Equivalence Ratio Concept and the Formation Mechanisms of Carbon-Monoxide in Enclosure Fires*, Progress in Energy and Combustion Science, 1995, **21 (3)** p. 197-237.

19. Beyler, C. L., *Fire Safety Science--Proceedings of the First International Symposium*, Hemisphere: New York (1991) 431.

20. Snegirev, A. Y., Makhviladze, G. M., Talalov, V. A. and Shamshin, A. V., *Turbulent Diffusion Combustion under Conditions of Limited Ventilation: Flame Projection Through an Opening*, Combustion, Explosion and Shock Waves, 2003, **39 (1)** p. 1-10.

21. Bertin, G., Most, J. M. and Coutin, M., *Wall fire behavior in an under-ventilated room*, Fire Safety Journal, 2002, **37 (7)** p. 615-630.

22. Utiskul, Y., Quintiere, J. G., Rangwala, A. S., Ringwelski, B. A., Wakatsuki, K. and Naruse, T., *Compartment fire phenomena under limited ventilation*, Fire Safety Journal, 2005, **40 (4)** p. 367-390.

23. Yii, E. H., Buchanan, A. H. and Fleischmann, C. M., *Simulating the effects of fuel type and geometry on post-flashover fire temperatures*, Fire Safety Journal, 2006, **41 (1)** p. 62-75.

24. Yii, E. H., Fleischmann, C. M. and Buchanan, A. H., *Experimental study of fire compartment with door opening and roof opening*, Fire and Materials, 2005, **29 (5)** p. 315-334.

25. National Institute of Standards and Technology, *Final report on the collapse of the World Trade Center towers*, 2005, **NIST NCSTAR 1**.

26. Grosshandler, W. L., Bryner, N. P., Madrzykowski, D. and Kuntz, K., *Report of the technical investigation of The Station nightclub fire*, 2005, **NIST NCSTAR 2**.

27. Madrzykowski, D. and Walton, W. D., *Cook County Administration Building fire, 69 West Washington, Chicago, Illinois, October 17*, 2004, **NIST SP1021**.

28. Hamins, A., Maranghides, A., Johnson, E. L., Donnelly, M. K., Yang, J. C., Mulholland, G. W. and Anleitner, R., *Report of Experimental Results for the International Fire Model Benchmarking and Validation Exercise #3*, 2005, **NIST SP1013-1**.

29. Huggett, C., *Estimation of Rate of Heat Release by Means of Oxygen-Consumption Measurements*, Fire and Materials, 1980, **4 (2)** p. 61-65.

30. Parker, W. J., *Calculations of the Heat Release Rate by Oxygen-Consumption for Various Applications*, Journal of Fire Sciences, 1984, **2 (5)** p. 380-395.

31. Bryant, R. A., Ohlemiller, T. J., Johnsson, E. L., Hamins, A., Grove, B. S., Maranghides, A., Mulholland, G. W. and Guthrie, W. F., *The NIST 3 Megawatt Quantitative Heat Release Rate Facility - Description and Proceedure*, 2004, **NIST IR 7052**.

32. Blevins, L. G. and Pitts, W. M., *Modeling of bare and aspirated thermocouples in compartment fires*, Fire Safety Journal, 1999, **33 (4)** p. 239-259.

33. Bryant, R., Womeldorf, C., Johnsson, E. and Ohlemiller, T., *Radiative heat flux measurement uncertainty*, Fire and Materials, 2003, **27 (5)** p. 209-222.

34. Pitts, W. M., Murthy, A. V., de Ris, J. L., Filtz, J. R., Nygard, K., Smith, D. and Wetterlund, I., *Round robin study of total heat flux gauge calibration at fire laboratories*, Fire Safety Journal, 2006, **41 (6)** p. 459-475.

35. Velmex, *XSLIDE Slide Assembly*, 2011, http://www.velmex.com/pdf/other/xslide.pdf.

36. Taylor, B. N. and Kuyatt, C. E., *Guidelines for Evaluating and Expressing the Uncertainty of NIST Measurement Results*, 1994, **NIST TN 1297**.

37. Bilger, R. W., *Reaction-Rates in Diffusion Flames*, Combustion and Flame, 1977, **30 (3)** p. 277-284.

38. Peters, N., *Laminar Diffusion Flamelet Models in Non-Premixed Turbulent Combustion*, Progress in Energy and Combustion Science, 1984, **10 (3)** p. 319-339.

39. Hamins, A. and Seshadri, K., *The Structure of Diffusion Flames Burning Pure, Binary, and Ternary Solutions of Methanol, Hepatane, and Toluene*, Combustion and Flame, 1987, **68 (3)** p. 295-307.

40. Sivathanu, Y. R. and Faeth, G. M., *Generalized State Relationships for Scalar Properties in Nonpremixed Hydrocarbon Air Flames*, Combustion and Flame, 1990, **82 (2)** p. 211-230.

41. Floyd, J. E., Wieczorek, C. J. and Vandsburger, U., *Simulation of the Virginia Tech Fire Research Laboratory Using Large Eddy-Simulation With Mixture Fraction Chemistry and*

Finite Volume Radiative Heat Transfer, Proceedings of the 9th International Interflam Conference, Volume 1. September 17-19, 2001, Edinburgh, Scotland, 2001, p. 767-778.

42. Pitts, W. M., Bryner, N. P., and Johnson, E. L., *Combustion Product Formation in Under and Overventilated Full-Scale Enclosure Fires,*, Combustion Institute/Central and Western States (USA) and Combustion Institute/Mexican National Section and American Flame Research Committee. Combustion Fundamentals and Applications. Joint Technical Meeting, San Antonio, TX, p. 565-570.

43. Sivathanu, Y. R. and Faeth, G. M., *Temperature Soot Volume Fraction Correlations in the Fuel-Rich Region of Buoyant Turbulent-Diffusion Flames*, Combustion and Flame, 1990, **81 (2)** p. 150-165.

44. Köylü, U. O. and Faeth, G. M., *Carbon-Monoxide and Soot Emissions from Liquid-Fueled Buoyant Turbulent-Diffusion Flames*, Combustion and Flame, 1991, **87 (1)** p. 61-76.

45. Puri, R. and Santoro, R. J., Proc.3rd Int.Sym.Fire Safety Science, 1991, p. 595-604.

46. G.H.Ko, A.Hamins, E.L.Johnson, M.Bundy, S.C.Kim and D.B.Lenhert, *Mixture fraction analysis of combustion products in the upper layer of reduced-scale compartment fires*, Combustion and Flame, 2009, **156 (2)** p. 467-476.

47. Quintiere, J. G., *Scaling Applications in Fire Research*, Fire Safety Journal, 1989, **15 (1)** p. 3-29.

APPENDICES

A. CHANNEL LISTS

LFL MIDAS Hookup Sheet Instrument and Channel Description	x x₀= inside south wall	y y₀= inside east wall	z z₀= floor (top of insulation)	file =	Location Description (doorway faces east)	Series: Underventilated ISO9705 Room	Overall Channel Number	Abbr.	MIDAS Station	Mod.	Mod. Ch. No.	Conv. Units	Wire	Gain	Grnd?	Revision Date: 4/8/2009
Measurements																
Main Channels	FSEmax=(240,360,240															
O2 at FRONT Sampling location (Rack #1)	120	85	Z		25 cm from Top, 50 cm from Right, 70 cm from Rear		0	O2Front	Center	1	0	Vol Fr	Cu	1		
Tamb					At MIDAS Center station		1	Tamb	Center	1	1	°C	TC	100		
CO2 at FRONT (Rack1) Sampling location (CO signal) overrange	120	85	Z		25 cm from Top, 50 cm from Right, 70 cm from Rear		2	CO2Front	Center	1	2	Vol Fr	Cu	1		
CO at FRONT (Rack1) Sampling location	120	85	Z		25 cm from Top, 50 cm from Right, 70 cm from Rear		3	COFront	Center	1	3	Vol Fr	Cu	1		
UH at FRONT (Rack1) Sampling location (before GC)	120	85	Z		25 cm from Top, 50 cm from Right, 70 cm from Rear		4	UHFront	Center	1	4	Vol Fr	Cu	1		
FRONT (Rack1) Gas Rack Dewpoint					In Rack #1 after Gas Dryer		5	DPFront	Center	1	5	°C	Cu	1		
FRONT (Rack1) Sampling location gas temperature 1-near probe					outside RSE after water cooled probe		6	TGasFront1	Center	1	6	°C	TC	100		
FRONT (Rack1) Sampling location gas temperature 2-near Dewpoint					In Rack #2 after dewpoint sample cell		7	TGasFront2	Center	1	7	°C	TC	100		
FRONT (Rack1) Temperature at Sampling Location Aspirated Type K	120	85	Z		25 cm from Top, 25 cm from Right, 70 cm from Rear		8	TFSamp	Center	1	8	°C	TC	100		
FRONT (Rack1) Sampling location water outlet temperature					outside room on water cooled probe		9	TH2OOutFront	Center	1	9	°C	TC	100		
O2 at Center (Rack2) Sampling location (Rack #2)	120		Z		25 cm from C, 50 cm from Right, 25 cm from Front		10	O2Center	Center	1	10	Vol Fr	Cu	1		
CO2 at Center (Rack2) Sampling location	120		Z		25 cm from C, 50 cm from Right, 25 cm from Front		11	CO2Center	Center	1	11	Vol Fr	Cu	1		
CO at Center (Rack2) Sampling location	120		Z		25 cm from C, 50 cm from Right, 25 cm from Front		12	COCenter	Center	1	12	Vol Fr	Cu	1		
UH at Center (Rack2) Sampling location (before auto sample storage)	120		Z		25 cm from C, 50 cm from Right, 25 cm from Front		13	UHCenter	Center	1	13	Vol Fr	Cu	1		
Center (Rack2) Gas Rack Dewpoint					In Rack #1 after Gas Dryer		14	DPCenter	Center	1	14	°C	Cu	1		
Center (Rack2) Sampling location gas temperature 1- near probe					outside room after water cooled probe		15	TGasCenter1	Center	1	15	°C	TC	100		
Center (Rack2) Sampling location gas temperature 2- near Dewpoint					In Rack #1 after dewpoint sample cell		16	TGasCenter2	Center	1	16	°C	TC	100		
Center (Rack2) Temperature at Sampling Location Aspirated Type K	120		Z		25 cm from C, 25 cm from Right, 25 cm from Front		17	TCSamp	Center	1	17	°C	TC	100		
Center (Rack2) Sampling location water outlet temperature					outside room on water cooled probe		18	TH2OOurCenter	Center	1	18	°C	TC	100		
O2 at REAR (Rack3) Sampling location (Rack #3)	120		Z				19	O2Rear	Center	1	19	Vol Fr	Cu	1		
CO2 at REAR (Rack3) Sampling location	120		Z				20	CO2Rear	Center	1	20	Vol Fr	Cu	1		
CO at REAR (Rack3) Sampling location	120		Z				21	CORear	Center	1	21	Vol Fr	Cu	1		
UH at REAR (Rack3)g Sampling location (before auto sample storage)	120		Z				22	UHRear	Center	1	22	Vol Fr	Cu	1		
REAR (Rack3) Gas Rack Dewpoint	120		Z				23	DPRear	Center	1	23	°C	Cu	1		
REAR (Rack3) Sampling location gas temperature 1- near probe	120		Z		outside room after probe		24	TGasRear1	Center	1	24	°C	TC	100		
REAR (Rack3) Sampling location gas temperature 2- near Dewpoint	120		Z				25	TGasRear2	Center	1	25	°C	TC	100		
REAR (Rack3) Sample in room Aspirated Type K	120		Z		moving		26	TRSamp	Center	1	26	°C	TC	100		
Textra							27	Textra	Center	1	27	°C	TC	100		
New Soot Probe Sample Temperature Outer							28	TStSampROut	Center	1	28	°C	TC	100		
Gravimetric Soot Probe 1/3 Mass Flow -rear					outside RSE		29	SootMF1	Center	1	29	V	Cu	1		
Gravimetric Soot Probe 2/4 Mass Flow-front					outside RSE		30	SootMF2	Center	1	30	V	Cu	1		
Gravimetric Soot Probe 3 Filter Temperature - rear					outside RSE		31	TGravSoot1	Center	1	31	°C	TC	100		
Gravimetric Soot Probe 4 Filter Temperature -rear					outside RSE		32	TGravSoot2	Center	2	0	°C	TC	100		
Gravimetric Soot Probe 3 Solenoid Signal -rear New Probe					outside RSE		33	VGravSoot1	Center	2	1	V	Cu	1		
Gravimetric Soot Probe 4 Solenoid Signal -rear New Probe Gravimetric					outside RSE		34	VGravSoot2	Center	2	2	V	Cu	1		
Gravimetric Soot Probe 1 Filter Temperature -front					outside RSE		35	TGravSoot3	Center	2	3	°C	TC	100		
Gravimetric Soot Probe 2 Filter Temperature-front					outside RSE		36	TGravSoot4	Center	2	4	°C	TC	100		
Gravimetric Soot Probe 1 Solenoid Signal-front					outside RSE		37	VGravSoot1F	Center	2	5	V	Cu	1		
Gravimetric Soot Probe 2 Solenoid Signal-front					outside RSE		38	VGravSoot2F	Center	2	6	V	Cu	1		
Total Heat Flux Gauge Rear Floor A (SN=131835)	119.5	266	0		In floor on room CL, 90 cm from Rear		39	HFRFL	Center	2	7	kW/m2	Cu	100		
Total Heat Flux Gauge Front Floor B (SN=131836)	119.5	90	0		In floor on room CL, 90 cm from Front		40	HFFFL	Center	2	8	kW/m2	Cu	100		
Total Heat Flux Gauge Outside Floor C (SN=131833)	119.5	-20	0		Outside floor, -X cm from front		41	HFOFL	Center	2	9	kW/m2	Cu	100		
Total Heat Flux Gauge Rear Ceiling D (SN=131837)	119.5	266	233		In ceiling in room CL, 90 cm from Rear		42	HFRCE	Center	2	10	kW/m2	Cu	100		
Total Heat Flux Gauge Center Ceiling E (SN=131838)	119.5	178	233		In ceiling in room CL,centered		43	HFCCE	Center	2	11	kW/m2	Cu	100		
Total Heat Flux Gauge Front Ceiling F (SN=131834)	119.5	90	233		In ceiling in room CL, 90 cm from Front		44	HFFCE	Center	2	12	kW/m2	Cu	100		
Temperature of Total Heat Flux Gauge Rear Floor A	119.5	266	0		In floor on room CL, 90 cm from Rear		45	THFRFL	Center	2	13	°C	TC	100		
Temperature of Total Heat Flux Gauge Front Floor B	119.5	90	0		In floor on room CL, 90 cm from Front		46	THFFFL	Center	2	14	°C	TC	100		
Temperature of Total Heat Flux Gauge Outside Floor C	119.5	-20	0		Outside floor, -X cm from front		47	THFOFL	Center	2	15	°C	TC	100		
Temperature of Total Heat Flux Gauge Rear Ceiling D	119.5	266	233		In ceiling in room CL, 90 cm from Rear		48	THFRCE	Center	2	16	°C	TC	100		
Temperature of Total Heat Flux Gauge Center Ceiling E	119.5	178	233		In ceiling in room CL,centered		49	THFCCE	Center	2	17	°C	TC	100		
Temperature of Total Heat Flux Gauge Front Ceiling F	119.5	90	233		In ceiling in room CL, 90 cm from Front		50	THFFCE	Center	2	18	°C	TC	100		
Interior Enclosure Surface Temp Near Total HF Gauge Rear Floor A					A17		51	TSHFRFL	Center	2	19	°C	TC	100		
Interior Enclosure Surface Temp Near Total HF Gauge Front Floor B					A19		52	TSHFFFL	Center	2	20	°C	TC	100		
Interior Enclosure Surface Temp Near Total HF Gauge Outside Floor C							53	TSHFOFL	Center	2	21	°C	TC	100		
Interior Enclosure Surface Temp Near Total HF Gauge Rear Ceiling D					A11		54	TSHFRCE	Center	2	22	°C	TC	100		
Interior Enclosure Surface Temp Near Total HF Gauge Center Ceiling E					A13		55	TSHFCCE	Center	2	23	°C	TC	100		
Interior Enclosure Surface Temp Near Total HF Gauge Front Ceiling F					A15		56	TSHFFCE	Center	2	24	°C	TC	100		
Exterior Enclosure Surface Temp Near Total HF Gauge Rear Floor A					A18		57	TSXHFRFL	Center	2	25	°C	TC	100		
Exterior Enclosure Surface Temp Near Total HF Gauge Front Floor B					A20		58	TSXHFFFL	Center	2	26	°C	TC	100		
Exterior Enclosure Surface Temp Near Total HF Gauge Rear Ceiling C					A12		59	TSXHFRCE	Center	2	27	°C	TC	100		
Exterior Enclosure Surface Temp Near Total HF Gauge Center Ceiling D					A14		60	TSXHFCCE	Center	2	28	°C	TC	100		
Exterior Enclosure Surface Temp Near Total HF Gauge Front Ceiling E					A16		61	TSXHFFCE	Center	2	29	°C	TC	100		
Load Cell #1 300 kg for single burner and 1st of 2 burners							62	Mass1	Center	2	30	kg	Cu	1		
Load Cell #2 150 kg for 2nd of 2 burners							63	Mass2	Center	2	31	kg	Cu	1		
Textra							64	Textra	Center	3	0	°C	TC	100		
Rear Temp at OLD Geometric Sampling Location Bare Bead							65	TRSampOldBB	Center	3	1	°C	TC	100		
Interior Surface Temp Side Wall Adjacent to Rear Sampling Position					A1		66	TSRSamp	Center	3	2	°C	TC	100		

LFL MIDAS Hookup Sheet Instrument and Channel Description				file =	Series:	Underventilated ISO9705 Room			Revision Date:			4/8/2009	
Measurements	x x₀= inside south wall	y y₀= inside east wall	z z₀= floor (top of insulation)	Location Description (doorway faces east)	Overall Channel Number	Abbr.	MIDAS Station	Mod.	Mod. Ch. No.	Conv. Units	Wire	Gain	Grnd?
Main Channels	FSEmax=(240,360,240)												
Interior Surface Temp Side Wall Adjacent to Rear Sampling Position				A1	66	TSRSamp	Center	3	2	°C	TC	100	
Exterior Surface Temp Side Wall Adjacent to Rear Sampling Position				A2	67	TSXRSamp	Center	3	3	°C	TC	100	
Plate Thermometer Back Surface Temperature					68	TDPTB	Center	3	4	°C	TC	100	
Plate Thermometer Front Surface Temperature					69	TDPTF	Center	3	5	°C	TC	100	
Textra					70	Textra	Center	3	6	°C	TC	100	
Surface Temp Near Test HF Gauge Center Ceiling E Pith Bare Bead				A3	71	TSHFCCEPtRh	Center	3	7	°C	TC	100	
Interior Surface Temp Side Wall Adjacent to Front Sampling Position				A4	72	TSFSamp	Center	3	8	°C	TC	100	
Exterior Surface Temp Side Wall Adjacent to Front Sampling Position				A4	73	TSXFSamp	Center	3	9	°C	TC	100	
Liquid Burner Fuel Temperature Low Positon	119.5	178			74	TFuelLow	Center	3	10	°C	TC	100	
Liquid Burner Fuel Temperature Mid Position Center	119.5	178		Connected to base of burner for pool height B14	75	TFuelMidC	Center	3	11	cm	Cu	1	
Liquid Burner Fuel Temperature Mid Position North					76	TFuelMidN	Center	3	12	°C	TC	100	
Liquid Burner Fuel Temperature Mid Position South					77	TFuelMidS	Center	3	13	°C	TC	100	
TC Tree Rear TC 1 in up	120	288	2.5	Rear 1 in up	78	TR3	Center	3	14	°C	TC	100	
TC Tree Rear TC 1 ft up	120	288	30	Rear 1 ft up	79	TR30	Center	3	15	°C	TC	100	
TC Tree Rear TC 2 ft up	120	288	60	Rear 2 ft up	80	TR60	Center	3	16	°C	TC	100	
TC Tree Rear TC 3 ft up	120	288	90	Rear 3 ft up	81	TR90	Center	3	17	°C	TC	100	
TC Tree Rear TC 3.5 ft up	120	288	105	Rear 3.5 ft up	82	TR105	Center	3	18	°C	TC	100	
TC Tree Rear TC 4 ft up	120	288	120	Rear 4 ft up	83	TR120	Center	3	19	°C	TC	100	
TC Tree Rear TC 4.5 ft up	120	288	135	Rear 4.5 ft up	84	TR135	Center	3	20	°C	TC	100	
TC Tree Rear TC 5 ft up	120	288	150	Rear 5 ft up	85	TR150	Center	3	21	°C	TC	100	
TC Tree Rear TC 6 ft up	120	288	180	Rear 6 ft up	86	TR180	Center	3	22	°C	TC	100	
TC Tree Rear TC 7 ft up	120	288	210	Rear 7 ft up	87	TR210	Center	3	23	°C	TC	100	
TC Tree Rear TC 7 ft 11 in up	120	288	237.5	Rear 7 ft 11 in up	88	TR237	Center	3	24	°C	TC	100	
TC Tree Front TC 1 in up	120	72	2.5	Front 1 in up	89	TF3	Center	3	25	°C	TC	100	
TC Tree Front TC 1 ft up	120	72	30	Front 1 ft up	90	TF30	Center	3	26	°C	TC	100	
TC Tree Front TC 2 ft up	120	72	60	Front 2 ft up	91	TF60	Center	3	27	°C	TC	100	
TC Tree Front TC 3 ft up	120	72	90	Front 3 ft up	92	TF90	Center	3	28	°C	TC	100	
TC Tree Front TC 3.5 ft up	120	72	105	Front 3.5 ft up	93	TF105	Center	3	29	°C	TC	100	
TC Tree Front TC 4 ft up	120	72	120	Front 4 ft up	94	TF120	Center	3	30	°C	TC	100	
TC Tree Front TC 4.5 ft up	120	72	135	Front 4.5 ft up	95	TF135	Center	3	31	°C	TC	100	
TC Tree Rear TC 5 ft up	120	288	150	Rear 5 ft up	85	TR150	Center	3	21	°C	TC	100	
TC Tree Rear TC 6 ft up	120	288	180	Rear 6 ft up	86	TR180	Center	3	22	°C	TC	100	
TC Tree Rear TC 7 ft up	120	288	210	Rear 7 ft up	87	TR210	Center	3	23	°C	TC	100	
TC Tree Rear TC 7 ft 11 in up	120	288	237.5	Rear 7 ft 11 in up	88	TR237	Center	3	24	°C	TC	100	
TC Tree Front TC 1 in up	120	72	2.5	Front 1 in up	89	TF3	Center	3	25	°C	TC	100	
TC Tree Front TC 1 ft up	120	72	30	Front 1 ft up	90	TF30	Center	3	26	°C	TC	100	
TC Tree Front TC 2 ft up	120	72	60	Front 2 ft up	91	TF60	Center	3	27	°C	TC	100	
TC Tree Front TC 3 ft up	120	72	90	Front 3 ft up	92	TF90	Center	3	28	°C	TC	100	
TC Tree Front TC 3.5 ft up	120	72	105	Front 3.5 ft up	93	TF105	Center	3	29	°C	TC	100	
TC Tree Front TC 4 ft up	120	72	120	Front 4 ft up	94	TF120	Center	3	30	°C	TC	100	
TC Tree Front TC 4.5 ft up	120	72	135	Front 4.5 ft up	95	TF135	Center	3	31	°C	TC	100	
TC Tree Front TC 5 ft up	120	72	150	Front 5 ft up	96	TF150	Center	4	0	°C	TC	100	
TC Tree Front TC 6 ft up	120	72	180	Front 6 ft up	97	TF180	Center	4	1	°C	TC	100	
TC Tree Front TC 7 ft up	120	72	210	Front 7 ft up	98	TF210	Center	4	2	°C	TC	100	
TC Tree Front TC 7 ft 11 in up	120	72	237.5	Front 7 ft 11 in up	99	TF237	Center	4	3	°C	TC	100	
Interior Surface Temperature Back Wall Centerline Top	125	356	180	A5	100	TSBWCTop	Center	4	4	°C	TC	100	
Interior Surface Temperature Back Wall Centerline Middle	125	356	120	A7	101	TSBWCMid	Center	4	5	°C	TC	100	
Interior Surface Temperature Back Wall Centerline Bottom	125	356	60	A9	102	TSBWCBot	Center	4	6	°C	TC	100	
Mass Flow Controller Output Rack 1				inside rack #1	103	MFRack1	Center	4	7	V	Cu	1	
Mass Flow Controller Output Rack 2				inside rack #2	104	MFRack2	Center	4	8	V	Cu	1	
Mass Flow Controller Output Rack 3				inside rack #3	105	MFRack3	Center	4	9	V	Cu	1	
Exterior Surface Temperature Back Wall Centerline Top	125	360	180	A6	106	TSXBWCTop	Center	4	10	°C	TC	100	
Exterior Surface Temperature Back Wall Centerline Middle	125	360	120	A8	107	TSXBWCMid	Center	4	11	°C	TC	100	
Exterior Surface Temperature Back Wall Centerline Bottom	125	360	60	A10	108	TSXBWCBot	Center	4	12	°C	TC	100	
Vertical Stage Temperature 1					109	TVertStage1	Center	4	13	°C	TC	100	
Vertical Stage Temperature 2					110	TVertStage2	Center	4	14	°C	TC	100	
Vertical Stage Temperature 3					111	TVertStage3	Center	4	15	°C	TC	100	
New Soot Probe Laser Detector #1 (variation 0-10V)					112	SootDet1Var	Center	4	16	V	Cu	1	
New Soot Probe Laser Detector #2 (signal 0-10V)					113	SootDet2Sig	Center	4	17	V	Cu	1	
Soot Sample Temperature					114	TSootSamp	Center	4	18	°C	TC	100	
Soot Sample Downstream Temperature					115	TSootSampDwn	Center	4	19	°C	TC	100	
Optical Soot Mass Flow Controller					116	OptSootMFC	Center	4	20	V	Cu	1	
Optics Cooling Water Temperature					117	TOptCool	Center	4	21	°C	TC	100	
New Soot Probe Sample Temperature At Measurement					118	TSootSampR	Center	4	22	°C	TC	100	
Aspirated Pump 1 Pressure (TRSampA and TFSampA)				outside RSE	119	PPump1	Center	4	23	Pa	Cu	1	
Aspirated Pump 2 Pressure (TRTreeTopA)				outside RSE	120	PPump2	Center	4	24	Pa	Cu	1	
Doorway Pressure Top					121	PDTop	Center	4	25	Pa	Cu	1	

LFL MIDAS Hookup Sheet Instrument and Channel Description				file =	Series:	Underventilated ISO9705 Room			Revision Date:		4/8/2009		
Measurements	x x₀= inside south wall	y y₀= inside east wall	z z₀= floor (top of insulat ion)	Location Description (doorway faces east)	Overall Channel Number	Abbr.	MIDAS Station	Mod.	Mod. Ch. No.	Conv. Units	Wire	Gain	Grnd?
Main Channels	FSEmax=(240,360,240)												
Doorway Pressure Bottom					122	PDBot	Center	4	26	Pa	Cu	1	
Doorway Top Aspirated Temperature					123	TDTopA	Center	4	27	°C	TC	100	
Doorway Top Bare Bead BiDi Temperature					124	TDTopBiDiBB	Center	4	28	°C	TC	100	
Aspirated Pump 1 Temperature (TRSampA and TFSampA)				B1	125	TPump1	Center	4	29	°C	TC	100	
Aspirated Pump 2 Temperaturee (TRTreeTopA)				B2	126	TPump2	Center	4	30	°C	TC	100	
Aspirated Pump 3 Temperature					127	TPump3	Center	4	31	V	Cu	1	
Soot Optics Pressure 1					128	POpt1	Center	5	0	V	Cu	1	
Soot Optics Pressure 2					129	POpt2	Center	5	1	V	Cu	1	
Plate Thermometer #1 Front Ceiling					130	PTCeF	Center	5	2	°C	TC	100	
Plate Thermometer #2 Center Ceiling					131	PTCeC	Center	5	3	°C	TC	100	
Plate Thermometer #3 Rear Ceiling					132	PTCeR	Center	5	4	°C	TC	100	
Plate Thermometer #4 Outside Floor					133	PTFlO	Center	5	5	°C	TC	100	
Plate Thermometer #5 Front Floor					134	PTFlF	Center	5	6	°C	TC	100	
Plate Thermometer #6 Rear Floor					135	PTFlR	Center	5	7	°C	TC	100	
Vertical Position of Moving Probes					136	MovePos	Center	5	8	cm	Cu	1	
O2 #3					137	O2-3S	Center	5	9	Vol Fr	Cu	1	
O2 Range #3					138	O2-3R	Center	5	10	V	Cu	1	
CO2 #3					139	CO2-3S	Center	5	11	Vol Fr	Cu	1	
CO2 Range #3					140	CO2-3R	Center	5	12	V	Cu	1	
CO #3					141	CO-3S	Center	5	13	Vol Fr	Cu	1	
CO Range #3					142	CO-3R	Center	5	14	V	Cu	1	
O2 #4					143	O2-4S	Center	5	15	Vol Fr	Cu	1	
O2 Range #4					144	O2-4R	Center	5	16	V	Cu	1	
CO2 #4					145	CO2-4S	Center	5	17	Vol Fr	Cu	1	
CO2 Range #4					146	CO2-4R	Center	5	18	V	Cu	1	
CO #4					147	CO-4S	Center	5	19	Vol Fr	Cu	1	
CO Range #4					148	CO-4R	Center	5	20	V	Cu	1	
5 V Marker Channel					149	5VMarker	Center	5	21	V	Cu	1	
Created Channels													
Event Marker 1					150	Event1							
Event Marker 2					151	Event2							
Doorway Velocity Top					152	VDTop		B	0	m/s			
Doorway Velocity Bottom					153	VDBot		B	1	m/s			
Soot Ratio					154	SootRatio		B	2	kg/kg			
O2#3					155	O2-3		B	3	n/a			
CO2#3					156	CO2-3		B	4				
CO#3					157	CO-3		B	5				
O2#4					158	O2-4		B	6				
CO2#4					159	CO2-4		B	7				
CO#4					160	CO-4		B	8				
Soot Volume Fraction from Optical Measurement					161	Fv		C	0				

B. EQUIPMENT LIST

Description	Manufacturer	Model
Oxygen analyzer for HRR	Servomex	540A
CO_2/CO analyzer for HRR	Seimens	Ultramat 6
Total HC analyzer for HRR	Rosemount	400A
Mass flow controller for HRR	MKS	1179A53C
Dew Point Transmitter for HRR	Vaisala	DMT242
Sample dryer for HRR	PermaPure	PD-200T-72SS
Sample pump for HRR	Gast	MOA-P122-AA
Liquid fuel turbine flow meter	Exact Flow	
Natural gas flow meter	Instromet	IRMA 15M-125
Total heat flux gauge (1--6)	Medtherm	16-0.75-10-4-12-36-20679k
Oxygen Analyzer (Rear, Front, and Move)	Servomex	4100
CO_2/CO Analyzer (Rear, Front, and Move)	Seimens	Ultramat 6E
Total HC analyzer (Rear, Front, and Move)	Baseline-Mocon	8800 H
Dew point meter (Rear, Front, and Move)	Vaisala	DMT242
Mass flow controller (Rear, Front, and Move)	MKS	M100B53C
MFC power supply	MKS	247D
Micro Gas Chromatograph (Front and Rear)	Agilent	3000A
Pressure Transducer for Velocity	MKS	220DD
Flow meter (spot check flows)	Bios Dry Cal	DCLT 20K
Venturi pump for aspirated TCs and gas sample tests #1-6	Vaccon	FDF 200-ST4
Gas conditioning system (Rear, Front, and Move)	PermaPure	MG-2812
Soot sample MFC	MKS	M100B53CCS1BV
Soot sample MFC	MKS	M100B53CCS1BV
MFC power supply	MKS	247D
Soot sample filter	Pall	P5PJ047
Soot sample cleaning pad	Hoppe's	1203
Soot sample filter holder	Gelman Sciences	2220
Soot sample 3-way solenoid valves	Parker	04F30C2208AAF4C05
Soot sample pumps	Gast	MOA-P122-AA
Linear stage	Velmex	MB9090K2J-S9
Load Cell (center)	Mettler Toledo	Jagxtreme KC300
Load Cell (rear)	Mettler Toledo	Jaguar KC150
Laser (Soot Measurement Probe)	Thorlabs	CPS192
Laser Detector (Soot Measurement Probe)	Thorlabs	DET110
Ceramic Blanket Insulation	ETS Shaefer	K-Lite HTZ
Ceramic Blanket Retainers	Refactory Anchors	RA-38
Spray Nozzels	BETE	WL 1 and WL 1-1/2

INDEX

www.ingramcontent.com/pod-product-compliance
Lightning Source LLC
Chambersburg PA
CBHW081830170526
45167CB00007B/2775